THE REGISTER OF

BAPTISMS, MARRIAGES AND BURIALS

AT ST. MARGARET'S

TOPPESFIELD PARISH, ESSEX CO., ENGLAND

1559-1650

AND

SOME ACCOUNT OF THE PARISH

THE REGISTER OF

BAPTISMS, MARRIAGES AND BURIALS

AT ST. MARGARET'S

TOPPESFIELD PARISH, ESSEX CO., ENGLAND

1559-1650

AND

SOME ACCOUNT OF THE PARISH

BY

REV. H. B. BARNES. RECTOR

AND

PHILIP MORANT

ISBN: 978-93-5128-776-6(PB)

First Published in 1905
Indian Reprint in 2017

Published by

Kalpaz Publications
C-30, Satyawati Nagar,
Delhi – 110052
E-mail: kalpaz@hotmail.com
Website: kalpazpublications.com
Ph.: +91-9212142040

On the fly leaf of the Register appears the following:

When Advent Clime to take his time,
then out goes wedding tide,
Like Artillary, in Comes Hillary,
with weddings at his side.
Septuagint takes the next hint,
and bids them next adewe,
But Ester Mass, wth eight days pass,
thou mayst get wedd anewe.
Rogation did yt last forbid,
& bid thee pray instedd,
But Trinity gives liberty,
to make a marriag bedd.

Conjugium Adventus tollit Hilarius relaxat
Septuagesima vetat sed pascæ octava reducit
Rogamen vetitat, commendit trina potestas.

Advent wils the to conteine
But Hilary sets the free again,
Septuagesima saies the nay
But Eight from Easter saies thou may.
Rogation bids the yet to tarrie
But Trinity gives the leave to marrie.

Duodecim Impedimenta matrimonii juxta Canonistas scil:

1. Error. 2. conditio. 3. votum. 4. cognatio. 5. crimen.

6. Cultus disparitas. 7. vis. 8. ordo. 9. legamen. 10. honestas.

11. Si sis affinis. 12. si forte coire nequibis.

On the reverse of the fly leaf is written:

Toppsfield steple fell downe
July the forth day 1689
and five beles and the little
bel broke all to peeceis

BAPTISMS.

[1]
1559 Anne Humfrye the daughter of John Humfrye was baptized the 24th. day of februarie in the yeere of o^{r} L. God 1559.
John, s. Thomas Webbe, 12th. March.
1560 Joane, d. Richard Yeldam, 31 March.
Margarete, d. John Purkis, 31 March.
John, s. Thomas Pollard, 7 April.
Henrie, s. James Edward, 21 April.
Elizabeth, d. John Hwes, 26 May.
John, s. Henrie Reade, 9 June.
William, s. Thomas Cracherood, 16 June.
William, s. John Plomb, 28 Julie.
John, s. Henrie Thetford, 1st. November.
Thomas, s. Henrie Smith, 22 November.
Henrie, s. John Bust, 26 September.
Anne, d. Thomas Spiltimber, 22 Januarie.
Anne, d. Thomas Greene, 23 Februarie.
Henrie, s. John Humfrie, 2 March.
1561 William, s. William Edward, 13 April.
John, s. John Underod, 7 September.
Anne, d. Richarde Motte, 5 October.
Elizabeth, d. Thomas Cracherood, 2 November.
Henrie, s. Henrie Snellocke, 14 December.
Rose, d. Richard Gipps, 26 December.

NOTE: The folios of the original volume are indicated by the figures enclosed within brackets. The first baptism is copied in the exact form of the original entry. The following records of baptism are printed in condensed form, the unnecessary verbiage having been omitted.

Robert, s. Thomas Maryet, 18 Januarie.
William, s. William Odell, 1 Februarie.
Richard, s. William Bateman, 24 Februarie.
Nicholas, s. Nicholas Alowe, last day of Februarie.
Alice, d. Henrie Cante, 8 March.
Katherine, d. Richard Gipps, 16 March.
John, s. John Humfrie, 22 March.
1562 Thomas, s. Christopher Fiche, 31 March.
Alice, d. William Addams, 26 Julie.
Thomas, s. Henrie Thetford, 19 Julie.
Thomas, s. Richard Yeldam, 13 September.
Matthew, s. William Edward, 17 Januarie.
Ann, d. Michael Tongue, 17 Januarie.
Anne, d. John Hwes, 10 Februarie.
Alyce, d. Richard Mott, 6 March.
Elizabeth, d. Henrie Snellocke, 6 March.
Elizabeth, d. John Edward, 6 March.
[2] Barbara, d. Thomas Cracherood, 6 April.
1563 Edward, s. John Coosen, 20 May.
William, s. John Humfrie, 6 April.
Margarete Tittrell, 28th. August.
Anne Fitche, 28th August.
Thomas Greene, 28th August.
Alice Edward, 29th August.
Cicely Cooke, 29 September.
Elizabeth Underwodde, 15 October.
Henrie Smith, 30 October.
Margarete Adams, 14 November.
Joane Gridlye, 2 December.
Joane Thetforde, 22 Januarie.
Elizabeth Chote, 4 March.
Marie Powell, 4 March.
Pleasant Cirke, 23 March.
1564 Barbara, d. William Buttall, 23 April.
Richard, s. Thomas Spiltimber, last day of April.
Barbara, d. Michael Tongue, 28 May.
Thomas, s. John Plombe, 30 Julie.
ffrances, d. Thomas Cracherood, 17 August.
John, s. John Rizing, 24 September.
Richard, s. Richard Yeldam, 20 September.

Thomas, s. John Humfrie, 25 October.
Barbara, d. Thomas Maryet, 12 November.
Rose, d. Robert Edward, 10 December.
Henrie, s. Henrie Reade, 28 December.
Anne, d. Thomas Gridlie, 1 Januarie.
Anne Hibys, d. Thomas Hibys, 5 Februarie.
ffrancis, s. John Auger, 24 Februarie.
Thomas. s. Thomas Pollard 11 March.
1565 Margerie, d. Thomas Webb, 8 April.
Elizabeth, d. William Edward, 18 April.
Henrie, s. Henrie Harrington, 19th May.
Margarete, d. William Tailor, 20 May.
John, s. Thomas Adcocke, 10th June.
Elizabeth, d. Richard Gipps, 10 June.
Simon, s. Thomas Greene, 20 June.
Thomas, s. John Coozin, 28 Julie.
ffrances, d. John Hwes, 28 Julie.
Robert, s. Thomas Edward, 19 August.
[3] Margarete, d. John Underwode, 19 August.
Anne Mott, d. William Mott, 29 September.
John, s. Christopher Fitch, 4 October.
Elizabeth, d. Steven Tytrill, 7 October.
Henry, s. Richard Mott, 14 October.
Richard, s. Richard Yeldam, 11 November.
Anne, d. Henrie Snellocke, 25 November.
William, s. Michael Tongue, 21 December.
Henrie, s. Henrie Thetford, 13 Januarie.
ffrances, d. John Plomb, 14 Januarie.
Richard, s. Henrie Smithe, 8 March.
Alyce, d. John Cirke, 11 March.
1566 William, s. Robert Edward, last day of March, 1566.
William, s. Thomas Mante, 2 April.
Elizabeth, d. Hugh Rawlin, 16 June.
William, s. William Butcher, 21 Julie.
Alice, d. William Buttell, 13 September.
Thomas, s. John Plomb, 10 November.
Marie, d. William Taylor, 10 November.
Barbara, d. Thomas Spiltimber, 28 December.
Rose, d. William Edward, 9 Februarie.
Anne, d. Ellen Peacock, 9th March, 1566, baseborn.

Barbara, d. John Humfrie, 18 March.
Elizabeth, d. John Coozin, 20 March.
1567 Richard, s. Steven Tittrill, 30 March.
Agnes, d. Richard Yeldam, 10 April.
Katherine, d. Thomas Hybys, 13 April.
William, s. William Browne, 20 April.
Alyce and ffrances, daughters of Michael Tongue, 24 April.
Joane, d. William Butcher, 15 June.
Briggite, d. Hugh Rawling, 16 Julie.
Richard, s. John Underwood, 3 August.
Elizabeth, d. Richard Perrie, 3 August.
Nicholas, s. Henrie Reade, 4 August.
Elizabeth, d. John Auger, 6 August.
Alyce, d. Henrie Snellock, 7 August.
Elizabeth, d. Thomas Pollard, 24 Sept.
Anne, d. John Plomb, 19 October.
Elizabeth & Jone, daughters of Thomas Adcocke, 2 November.
Richard, s. Christopher ffitch, 18 November.
Elizabeth, d. Henrie Smith, 1 Februarie.
Richard, s. Henrie Thetford, 15 Februarie.
Joane, d. Richard Hulle, 17 Februarie.
John, s. William Mott, 11 March.

[4]

1568 Barbara, d. Thomas Greene, 7th. April.
Cicely, d. Richard Motte, last day of April.
Margerie, d. John Mortemer, 2 Julie.
William, s. William Butcher, 5 August.
William, s. William Buttall, 3 October.
William, s. Richard Yeldam, 3 October.
Marie, d. Nicholas Waede, 24 October.
Agnes, d. John Underwode, 25 October.
Christopher, s. Henrie Snellocke, 15 Januarie.
Thomas, s. Hugh Rawlings, 6 Februarie.
William, s. Thomas Hybys, 24 Februarie.
Rose, d. John Cirke, 27 Februarie.
Henrie, s. Henrie Biggs, 13 March.
1569 Richard, s. William Edward, 28 March.
Elizabeth, d. Henrie Thetford, 29 May.

Ellin, d. John Humfrie, 12 June.
John, s. John Harrington, 27 Julie.
Thomas, s. Robert Edward, 2 August.
Henrie, s. Thomas Spiltymber, 2 October.
Rosanna, d. Edward Richardson, 25 November.
Ellin, d. William Tongue, 22 Januarie.
Elizabeth, d. Thomas Maye, 1 Februarie.
Margaret, d. Thomas Pollerd, 19 Februarie.
William, s. Thomas Greene, 6 March.
1570 William, s. Richard Perrye, 25 March.
Margarie, d. Richard Gipps, 25 April.
Edward, s. Robert Turner, last day April.
Elizabeth, d. saide Robert Turner, last day April.
ffrancis, s. Henrie Snellocke, 26 June.
Emily, d. Rose Plomb, widowe, 8th Julie.
Anne, d. Henrie Smith, 13 August.
Alice, d. John Underwode, 8 October.
Richard, s. Richard Motte, 12 November.
Samuel, s. William Edward, 25 December.
Edward, s. Christopher Fitch, 26 December.
Joyce, d. saide Christopher Fitch, 26 December.
Matthew, s. William Butcher, of Gaynsfords, 1st Januarie.
John, s. John Auger, 14 Januarie.
1571 Henrie, s. Thomas Maye, 6 April.
Alice, d. William Butcher, the thetcher, 6 May.
Marie, d. Henrie Thetford, 20 June.
Robert, s. Robert Aldreade, 12 August.
[5] John, s. Robert Edward, 25 August.
Edward, s. John Humfrie, 2 September.
William, s. Hugh Rawlinge, 7 September.
Joane, d. John Cirke, 18 November.
Edward, s. Edward Richardson, 14 December.
Dorcas, d. Henrie Snellocke, 14 Februarie.
1572 Thomas, s. Thomas Browne, 25 March.
Elizabeth, d. John Buttall, 20 April.
Rose, d. Thomas Spiltimber, 1 May.
Marie, d. Henrie Bigge, 10 August.
Joane, d. John Hamont, 19 September.
Dorcas, d. John Humfrie, 24 September.

Marie, d. William Buttall, 15 October.
Edward, s. Thomas Pollard, 4 Januarie.
Robert, s. William Edward, 11 Januarie.
Elizabeth, d. William Edward, 11 Januarie.
Robert, s. Robert Flowere, 18 Januarie.
Marie, d. Thomas Cracherood, sen., 10 Februarie.
1573 Margarie, d. Henrie Thetford, 25 March.
Phyllis, d. Richard Eeveryde.
Marie, d. Richard Motte, 30 August.
Susan, d. John Auger, 6 September.
Richard, s. Paul Rawlinge, 8 September.
ffrancis, s. Richard Yeldam, 13 September.
Dorcas, d. John Reade, 14 September.
John, s. Henrie Smith, 13 November.
Dorcas, d. William Butcher, of Gensforde, 22 November.
Richard, s. William Butcher, aforesaid, 22 November.
ffrancis, s. Henrie Billirod, otherwise called Bust, 29 November.
Robert, s. Robert Smith, 27 December.
William, s. Hugh Rawlinge, 26 December.
Wa[l]ter, s. Christopher Taylor, 27 December.
Samuel, s. Robert Edward, 5 Januarie.
Alice, d. Richard Gipps, the last Januarie.
[6] Anne, d. William Earelope, otherwise called Connye, 21 March.
1574 ffrances, d. John Hamonte, 24 April.
William, s. John Clarke, 13 June.
Margarete, d. Edward Richardson, 23 Julie.
Margaret, d. William Boram, 8 August.
Leastrange, s. Henrie Snellocke, 22 August.
Margarete, d. Richard Eeverede, 29 August.
Marie, d. Thomas Bailye, 21 November.
William, s. John Harvie, 14 December.
Marie, d. the aforesaide John Harvie, 14 December.
Anna, d. Thomas Cracherood the elder, 19 Januarie.
Henrie, s. John Cirke, 13 Februarie.
Marie, d. Henrie Wayte, 20 Februarie.
William, s. Steven Cante, 20 March.
1575 John, s. Richard Hulle, 26 April.

William, s. Robert fflowere, 1 May.
Elizabeth, d. Thomas Browne, 3 May.
Barbara, d. William Redman, 15 May.
Susan, d. John Humfrie, 15 May.
Thomas, s. Thomas Aldrede, 19 June.
Robert, s. William Reade, 26 Julie.
Anne, d. Thomas Pollard, 4 September.
William, s. Christopher Fitch, 18 September.
William, s. William Bigge, 26 September.
Margarete, d. Henrie Gridlie, 31 December.
Susan, d. John Buttall, 26 Februarie.
Hugh, s. Hugh Rawlinge, 18 Februarie.
Priscilla, d. William Buttall, 4 March.
1576 Alyce, d. Edward Richardson, 9 Julie.
ffrances, d. William Butcher, 15 Julie.
William, s. William Reade, 7 August.
ffrancis, s. Henrie Fetforth, 12 August.
Winnefrede, d. Henrie Snellocke, 16 September.
Millicent, d. William Redman, 23 September.
Alyce, d. William Edward, 13 November.
R* s. Richard Butcher, 19 Februarie.
Joane, d. John Bateman, 20 March.
Marie, d. Steven Cante, 20 March.

[7]
1577 Elizabeth, d. John Brine, 26 March.
William, s. Thomas Browne, 3 April.
ffrances, d. Robert Edward, 5 April.
William, s. Thomas Cracherood the younger, 28 Julie.
Alice, d. John Greene, 3 November.
——, s. Richard Bateman, 3 November.
Rose, d. Robert Perrye, 24 November.
——, d. John Buttall, 17 December.
John, s. Richard Evered, 22 December.
1578 Robert, s. John Clarke, 1 April.
Ralfe, s. John Humfrie, 6 April.
John, s. —— Horklye, 18 May.
Thomas, s. Thomas Cracherood, 15 June.
William, s. William Pollard, 18 November.

---

*Rest of name erased.

William, s. William Reade, 23 Julie.
Ralfe, s. William Reade, 23 Julie.
1579 Anne, d. Steven Cante, the last day of May.
Dorcas, d. William Redman, 18 June.
Henry Bateman was baptized 20 June.
Marie, d. Robert Edward, 13 August.
Alice, d. John Greene, 6 September.
Trefyna, d. William Boram, 13 September.
Annie, d. Richard Evered, 18 October.
Anne, d. Thomas Browne, 25 October.
John, s. Henrie Gridley, 22 November.
Richarde, s. Robert Smith, 22 November.
Margarete, d. Richard Perrye, 26 November.
Sarah, d. Richard Edward, 17 Januarie.
Thomas, s. William Buttall, 17 Januarie.
Elizabeth, d. William Pollard, 17 Januarie.
Henrie, s. John Bateman, 24 Januarie.
Matthew, s. Thomas Cracherood the younger, 24 Februarie.
John, s. John Bottall [*sic*], 24 Februarie.
1580 Henrie, s. John Clarke, 26 March.
Samuel, s. William Bigge, 15 May.
William, s. John Hammond, 15 June.
Audre, d. Richard Bocher, 15 June.
William, s. Edward Richardson, the last day of Julie.
Margarete, d. Robert George, 7 August.
Margarete, d. Thomas Bateman, 13 November.
Margarete, d. Thomas Garner, 20 November.
Thomas, s. William Rede, 5 March.
1581 Robert, s. Henrie Snellocke, 8 May.
Anne, d. William Harrington, 14 May.
[8] Sarah, d. John Briant, 21 May.
Henrie, s. Thomas Baylie, 2 Julie.
Jane, d. Thomas Cracherood the elder, 16 Julie.
William, s. Robert Joley, 1 September.
William, s. Egidii Rede, 29 October.
Sarah, d. William Bacon, 26 November.
Anne, d. Thomas Cracherood the younger, 21 Januarie.
Marie, d. Richard Bateman, 28 Januarie.
John, s. Robert Perry, 12 Februarie.

1582 Judeth, d. Steven Cante, 25 March.
Elizabeth, d. Edward Richardson, 1 April.
Margerie, d. Clement Boram, 1 April.
Henrie, s. Henrie Gridley, 8 April.
Christian, d. Robert Edward, 17 April.
William, s. Robert Overed, 17 April.
Barbara, d. Richard Evered, 22 April.
William, s. Thomas Bateman, 20 May.
John, s. Thomas Browne, 20 May.
Anne, d. Thomas Garner, 24 June.
Robert, s. Robert George, 1 Julie.
Elizabeth, d. Robert Greene, 14 Julie.
William, s. John Bosall, 28 Julie.
James, s. John Harrington, yeoman, 11 August.
John, s. John Harrington, paup., 11 August.
Judith, d. William Pollard, 9 September.
Joane, d. John Waford, 13 October.
Anne, d. John Hamond, 4 November.
Margarete, d. William Bosall, 20 Januarie.
Alyce, d. Roger Edward, 3 Februarie.
1583 John, s. John Freer, 7 April.
Dorothy, d. Egidii Rede. 7 April.
Elizabeth, d. John Greene, 26 Maye.
Thomas, s. Edward Laver, 2 June.
Ishmael, s. Margarete Clarke, adult, 13 June.
James, s. John Harrington, 29 September.
William, s. William Bacon, 6 October.
Briget, d. John Bryant, 13 October.
William, s. Richard Bucher, 27 October.
Richard, s. Richard Evered, 10 November.
Elizabeth, d. William Reade, 10 December.
Elizabeth, d. Robert Perry, 15 December.
Anne, d. John Wafer, 22 December.
William, s. William Cracherood the younger, 25 December.
Thomas, s. Nicholas Garnet, 15 Januarie.
[9] Jane, d. Thomas Cracherood the younger, 23 Januarie.
Marie, d. Thomas Carey, 2 Februarie.
Elizabeth, d. Thomas Bateman, 1 March.
Edmund, s. John Hamond, 15 March.

1584 Joyce, d. John Hampton, 3 May.
Samuel, s. Robert Hamond, 24 May.
Robert, s. Robert Edward, 8 June.
Alyce, d. Thomas Garner, 9 August.
Margarete, d. William Harrington, 29 August.
Joyce, d. Richard* Bateman, 29 August.
Edward, s. John Reade, 20 September.
Steven, s. Steven Cante, 4 October.
Henrie, s. Egidii Reades, 1 November.
Edward, s. William Bigge, 15 November.
James, s. Edward Richard, 15 November.
Penelope, d. Richard Evered, 13 December.
Samuel, s. Robert Overed, 17 Januarie.
Joane, d. Henrie Gridley, 17 Januarie.
Thomas and Robert, s. John Harrington, paup., 31 Januarie.
Anne, d. Henrie Bateman, 1 March.
1585 Alyce, d. John Bosall, 12 April.
Alyce, d. Robert Perrye, 1 May.
Marie, d. John Harrington, 2 May.
Andrew, s. Roger Edward, 2 May.
William, s. John Teball, 9 May.
John, s. John Greene, 9 May.
——, s. John Waford, 16 May.
William, s. William Edward, jun., 20 June.
Barbara, d. Henrie Snellocke the younger, 11 Julie.
Thomas, s. Robert George, 18 Julie.
Anne, d. John Sewell, 26 September.
Henrie, s. Elizabeth Chunke, widow [*sic*], 14 November.
Martha, born in fornication of Humfrye Elsworth and Anne Hewes, 15 Januarie.
1586 Sarah, d. Thomas Bateman, 4 April.
Dorcas, d. William Bigge, 7 April.
Leastrange, s. William Firmin, 7 April.
John, s. Thomas Cracherood, jun., 10 April.
Anne, d. William Buttall, 15 April [*sic*].
Nathan, s. Richard Bateman, jun., 1 May.
Marie, d. Lodovice Brett, 1 May.

*Henrie erased and Richard inserted.

Hester, d. Edward Tilbrok, 1 May.
John, s. John Bateman, 15 May.
Henrie, s. William Cracherood, 18 May.
Anne, d. John Briant, 23 May.
Daniel, s. William Home, 5 June.
[10] Margarete, d. Thomas Browne, 16 June.
Richard, s. Henrie Snellocke, 6 September.
Sarah, d. John Perrye, 2 October.
Ann, d. John Cob, 10 October.
Matthew, s. Henrie Bever, 4 December.
Josias, s. John Pollard, 5 December.
ffaith, d. Richard Bocher, 11 December.
John, s. John Read, 26 Februarie.
Robert, s. James Robinson, 12 March.
——, of Robert Linwood, 12 March.
1587 William, s. Clement Boram, 9 April.
Thomas, s. John Waford, 14 May.
William, s. William Linsye, 23 May.
——, William Ferrer, 23 May.
Henrie, s. John Reade, jun., 11 June.
Anne, d. Henrie Snellock, jun., 30 Julie.
John, s. —— Sache, 14 August.
Sarah, d. Steven Cante, 10 September.
John, s. John Hamond, 17 September.
Richard, s. Henrie Gridley, 24 September.
Richard, s. John Gipps, 24 September.
Brigitte, d. Thomas Cracherood, jun., 1 October.
Tamesine, d. William Cracherood, 1 October.
Jone, d. William Browne, 1 October.
Elizabeth, d. William Edward, 19 November.
Clemens Tyboll, 19 November.
Rose, d. Edward Richardson, 3 Januarie.
Richard Howborowe, 3 Januarie.
ffrancis, s. John Somes, 4 Februarie.
Elizabeth, d. Richard Bateman, jun., 18 Februarie.
1588 Sarah, d. William Edward, of Bradfield, 25 Julie.
Margarie, d. Edward Osteler, 25 August.
Helene, d. Robert Tiler, 21 September.
Daniel, s. Robert Perrye, 29 September.
Thomas, s. Roger Edward, 13 October.

Susan, d. Henrie Snellocke, jun., 27 October.
Mattew, s. —— Brette, 11 November.
Rebekah, d. John Bryant, 11 November.
Anne, d. John Hart, 2 Februarie.
John, s. John Pollard, 9 Februarie.
Henrie, s. Richard Overed, 2 March.
Elizabeth, d. James Russell, 9 March.
1589 Anne, d. William ffirmin, 22 April.
Anne, d. Richard ——, 1 May.
[11] John Mariver, 25 Maye.
Alice ffyche, 25 May.
Elizabeth, d. John Reade, 29 June.
Ralfe, s. William Boram, 6 Julie.
John, s. Thomas Harvie, 27 Julie.
Prudence, d. John Cosin, 10 August.
John, s. John Warman, 14 September.
Alice, d. William Linsye, 14 September.
Hellena, d. Edward Osteler, 21 September.
Humfrie, s. Thomas Cracherood, jun., 25 September.
Marie, d. William Cracherood, 29 September.
Henrie, s. Henrie Snellock, jun., 5 October.
Thomas, s Thomas Corke, 5 October.
Susan, d. John Bird, 26 October.
Elizabeth, d. John Gipps, 9 November.
Susan, d. Jo— Ray de Gosfield,* 11 November.
Susan, d. Robert Tiler, 21 December.
William, s. William Thorogood, 18 Januarie.
Cicely, d. Robert Some, 18 Januarie.
Susan, d. Richard Bocher, 1 Februarie.
John, s. Marie Tilbroke, widow, 2 Februarie.
Grace, d. Steven Cante, 16 Februarie.
Elizabeth, d. William Ferrer, 16 Februarie.
Marie, d. Thomas Garner, 1 March.
Marie, d. Anne Bocher, borne in fornication, 1 March.
Elizabeth Howborowe, 1 March.
1590 Jane, d. Nicholas Reade, 5 May.
Elizabeth, d. William Edwards de Bradfields, 9 June.
William, s. Henrie Gridley, 28 Julie.

*Gosfield is a neighboring village.

William, s. Thomas ffyche.
Orphane, s. —— Powle, 11 October.
Christopher, s. Henrie Snellocke, jun., 24 November.
Henrie, s. Davide Marner, 29 November.
Anne, d. John Cosin, 6 December.
Margarete, d. William Argent, 1 Januarie.
John, s. Lewis Brett, 9 Februarie.
Roda, d. William Thorogood, 15 Februarie.
Susan, d. William Edward, jun., 25 Februarie.
Mercie, d. John Pollard, last day of Februarie.
1591 Susan, d. Thomas Bateman, 28 March.
Thomas. s. Richard Evered, 18 April.
William, s. William Bust, 2 May.
Joane, d. Henrie Edward, 9 May.
Rose, d. Henrie Laver, 9 May.
Robert, s. Thomas Cracherood, jun., 16 May.
Susan, d. John Bryant, 16 May.
Edward, s. William Cracherood, 20 June.
Zacharias, s. —— Marner, 15 August.
Thomas, s. Thomas Plomb, 15 August.
[12] Alice, d. Edward Osteler, 12 September.
Sarah, d. Henrie Bret, 28 September.
Thomas, s. Thomas Harvie, 10 October.
Richard, s. Isaac Hart, 28 November.
Rose, d. Henrie Snellocke, 19 December.
Grissell, d. Robert Tiler, 9 Januarie.
Edward, s. Thomas Somes, 5 March.
Tamasin, d. John Gipps, 12 March.
Matthew and Katherine, chn. William Thorogood, 23 March.
1592 William, s. William Browne, 9 April.
Sarah, d. Thomas Edward, 4 June.
Rebekah, d. Nicholas Reade, 4 June.
Anne, d. Robert Rolfe, 18 June.
John, s. John Cosin, 18 June.
Henrie, s. John Reade, 25 June.
Edwards, s. Steven Cante, 20 August.
Thamesin, d. James Russell, 17 September.
Matthew, s. John Hamond, 1 October.
Thomas, s. Thomas Plomb, 29 October.

Anne, d. William Cosin, 29 October.
Ezichiel, s. Izhak Cornwell, 5 November.
Margaret, d. Thomas Lamson, 1 Januarie.
Henrie, s. Lewis Brette, 14 Januarie.
William, s. John Humfrye, 28 Januarie.
Elizabeth, d. —— Browne, 5 Februarie.
Matthew, s. Henrie Snellocke, 27 Februarie.
William, s. Richard Payne, 3 March.
1593 William, s. Edward Osteler, 1 April.
Susan and Margaret, ds. William Edward, 1 May.
John, s. William Bocher, 7 June.
Joane, d. Thomas Cracherood, 8 Julie.
ffrances, d. William Thorogood, 22 Julie.
Sarah, d. Robert Tyler, 25 Julie.
———, of John Reade, 23 September.
William, s. William ffirmin, 27 September.
William, s. Henrie Pettit, 21 October.
John, s. John Bryant, 4 November.
Matthew, s. Thomas Hurrell, 9 September.
Anne, d. Thomas Tonge, 6 Januarie.
Robert, s. Robert Marner, 10 Februarie.
[13] Susan, d. Richard Bateman, 17 Februarie.
1594 Robert, s. Robert Hogg, 21 April.
Joane, d. Lewis Turner, 12 May.
Winifrede, d. James Russell, 19 May.
Elizabeth, d. Lewis Brett, 26 May.
John, s. Robert Rolfe, 9 June.
Marie, d. Matthew Whiting, 18 Julie.
George, s. John Gipps, 28 Julie.
Margarete, d. Henrie Brette, 28 Julie.
ffrances, d. William Greene, 25 August.
Dorcas, d. William Cracherood, 2 October.
Joseph, s. Izhak Cornwell, 6 October.
William, s. Jeremie Turkill, 6 October.
Thomas, s. William Browne, 17 November.
Elizabeth, d. Thomas Horkley, 25 November.
William, s. Davide Marner, 5 Januarie.
Winifrede, d. James Russell, 28 Januarie.
John, s. John Clarke, 4 Februarie.
John, s. Edward Osteler, 2 March.

Elizabeth, d. Thomas Harvie, 20 March.
1595 Alyce, d. "of a certeine begger", 30 March.
Samuel, s. William Edward, 6 April.
Thomas, s. Thomas Marner, 25 Maye.
John, s. Simon Greene, 6 Julie.
Robert, s. Robert Edward, 9 Julie.
Thomas, s. John Pollard, 3 August.
———, of Brette, 10 August.
Elizabeth, of Robert Rolfe, 24 August.
Grace, d. Henry Snellock, 7 September.
William, s. William ffirmin, 7 September.
Alyce, d. John Amyce, 7 September.
Margarie, d. William Cracherood, 7 September.
John, s. John Playle, 13 November.
John, s. John Parker, 7 March.
1596 Thomas, s. Henry Pettitt, 25 March.
Sarah, d. Robert Tiler, 19 April.
———, of James Russell, 20 May.
Samuel, s. Robert Rolfe, 5 September.
Christopher, s. Christopher Snellock, 14 September.
William, s. William Edward of Bradfields, 7 November.
Hester, d. Richard Bateman, 13 November.
[14] John, s. John ffisher, 5 December.
Alce, d. William ffisher, 6 Januarie.
Torearie, d. Henrie Snellock, 16 Januarie.
John, s. John ffiche, 30 Januarie.
William, s. William Greene, 20 Februarie.
Thomas, s. Jonah Spiltimber, 20 Februarie.
Susan, d. Robert Hogg, 24 Februarie.
Sarah, d. Zacharie Smyth, 5 March.
1597 Joseph, s. —— Bragge, 10 April.
Samuel, s. Edward Osteler, 17 April.
Anne, d. William Bocher, 8 May.
Rachel, d. John Gipps, 8 May.
Robert, s. Thomas Edward, 12 September.
Timothye, s. —— Cob, 21 September.
Elizabeth, d. Richard Edward, 14 October.
Susan, d. William Birde, 17 November.
Thomas, s. John Platte, 19 November.
John, s. John ffisher, 20 Februarie.

Susan, d. Edmund Whiting, 20 Februarie.
Elizabeth, d. Thomas Plombe, 20 March.
1598 Anne, d. Richard Payne, 2 April.
William, s. Simon Greene, 23 April.
Helena, d. of Habel Laver, 30 April.
Emily, d. of Thomas Harvie, 24 May.
Dorothy, d. John Pollard, 28 May.
Susan, d. Richard Wilson, 27 August.
Robert, s. Robert Tiler, 28 August.
Margarete, d. John ffiche, 17 September.
William, s. William Greene, butcher, 1 October.
John, s. Robert Warner, 9 October.
William, s. William Bocher of Gensforde, 15 October.
Elizabeth, d. Edmund Bryant, 15 October.
Elizabeth, d. Henrie Pettit, 30 October.
Joane, d. Robert Rolfe, 24 December.
William, s. Richard Edward, 24 December.
1599 John, s. John Redgewell, 9 Aprill.
Steven, s. Christopher Reade, 25 April.
Dorcas, d. Henry Laver, 26 April.
William, s. John Clarke, 29 April.
Thomas, s. John Humfrye, 1 May.
Marie, d. Robert Hogge, 1 May.
Stephanus Reade.*
David, s. David Warner, 17 May.
Mary, d. John Fisher, 27 May.
Susanna, d. Edward Ostler, 27 May.
William, s. John Lawson, 7 August.
Richard, s. Richard Titirel, 14 August.
Robert, s. —— Sibley, 30 September.
William, s. William Butcher, alias Adams, 3 June.
Barbara, d. John Fimis, 11 February.

---

*A line is drawn through this name. The record is apparently a copy in the same hand to here, and this may be the signature of the copyist. The same hand apparently makes the next 11 entries, but there are irregularities unobservable before, and the ink is very bad, and faded. The ink becomes good, and the handwriting entirely changes its character, and becomes larger with the entry Thomas Greene, but changes back again with Margaret Pettitt (1 Nov. 1600). From that point forwards the changes in ink and handwriting are frequent, the different number of entries on a folio will give some idea of the variation in the size of the characters.

Mary, d. Jonas Spiltimber, 20 February.
Margery, d. Robert Tongue, 11 February.
Elizabeth, d. Thomas Chamberlain, 25 February.
Thomas, s. William Greene, 18 Februarie.
1600 Richard, s. William Greene, 30 March.
John, s. William Battie, 30 March.
Susanna, d. Gabriell Grante, 6 April.
Rachell, d. Abell Laver, 6 April.
Henry, s. Thomas Harvie, 20 April.
Thomasin, d. William Edward, 30 May.
Daniell, s. William Butcher, 30 May.
Sara, d. William Edward, jun., 17 August.
Winifrede, d. John Hayle, 31 August.
Rose, d. Richard Edward, 8 Septembe [*sic*].
Sara, d. Thomas Gow, 28 Septembe.
Hercules, s. widow ffisher, 28 Septembe.
Mary, d. Thomas Warner, 26 Octobe.
Thomazine, d. —— Chamberlin, 26 Octobe.
[15] Margaret, d. George Lane, 9 Novembe.
Mary, d. John Constable, 16 November.
Elizabeth, d. Richard Payne, 16 Novembe.
John, s. John Baylie, 23 Novembe.
Mary, d. John Ridgewell, 30 Novembe.
1601 Catheren, d. John Eoull, 1 November.
John, s. John Gipp, 1 November.
Margaret, d. William [erased] Pettitt, 1 November.
Thomas, s. Henry Cirke, 22 November.
1601 Henry, s. Henry Lande, 6 January.
1601 Mary, d. Richard Titrell, 6 January.
1601 Thomas, s. Thomas Sybley, 9 ffebruary.
1601 Agnes, d. William Butcher, 23 ffebruary.
1601 Robert, s. William Butcher, 7 March.
1602 Elizabeth, d. William Grene, 30 May.
1602 William, s. Mr. Robert Rolfe, 1 June.
Margaret, d. Adler Newman, 27 June [*sic*].
ffrancis, d. Michael Clerk, 8 August.
Robert, s. John Pollard, 8 August.
——, d. Richard Edwarde, 8 August.
Dorathy, d. Robt. Hoy, 6 September.
Peter, s. Davy Warner, 7 October.

William, s. William Battie, 7 October.
Edwards, s. Nicholas Evens, 7 October.
Agnes, d. Thomas Harvey, 19 October.
Alyce, d. Robert Spiltimber, 7 November.
Robert, s. Robert Clarke, 18 [?] November.
Ann, d. John Ridgewell, 23 January.
William, s. Thomas Plumb, 30 January.
Cicely, d. Edmund Briant, 9 ffebruary.
John, s. Samuel Edwarde, 8 March.

Jacobi primo.

1603 Mary, d. John Bayley, 5 May.
Ann, d. Robert Edwarde, 21 June.
Margaret, d. John Amis, 21 June.
Margery, d. John Start, 3 July.
Edwards, s. John Pitches, 17 July.
John, s. Richard Edward, 11 September.
ffrancis, s. Christopher Read, 14 September.
John, s. John Pettitt, 18 September.
Susan, d. John Pettitt, 18 September.
Mary, d. Richard Payne, 5 November.
Alyce, d. Richard Tittrell, 15 January.
Lucy, d. Thomas Sibble, 5 March.
Ann, d. William Butcher, 18 March.

Jacobi 2.

1604 John, s. Edmund Bryant, 20 April.
William, s. William Horner, 20 April.
John, s. John Perry, 22 May.
Ales, d. Adlan [*sic*] Newman, 9 Sept.
John, s. Richard Edwarde, 12 Sept.
Samuel, s. John Parker, 11 October.
Robert, s. Robert Spiltimber, 14 October.
Peter, s. Thomas Harvy, 28 October.
[16] Marian, d. Saunder Bulloynes, 11 November.
Henry, s. Atlas Evans, 11 November.
Susan, d. John Ridgwell, 18 November.
Priscilla, d. Michael Clerke, 13 January.
Dorathy, d. Thomas Warner, 20 January.
John, s. Henry Lawer, 7 ffebruary.
Ales, d. William Batte, 7 ffebruary.
William, s. John Gips, 6 March.

Jacobi 3.
1605 Mathews, s. Robert Edwarde, of ffullers, 25 March.
——, Thomas Plumbe, 7 Aprill.
Sara, d. James Shed, 5 May.
Sara, d. Thomas Chattowton, 19 May.
Henry, s. John Smith, 26 May.
Anne, d. William Greene, 13 June.
John, s. William Waford, 31 July.
Robert, s. Richard Edward, 29 August.
John, s. William Seman, 22 December.
Ann, d. John Amis, 26 December.
Robert, s. Edmund Briant, 1 January.
Richer, s. Ales Hart, 13 January.
William, s. Thomas Sibble, 4 February.
Jacobi quarto.
1606 John, s. John Fox, 30 March.
William, s. Thomas Chatterton, 1 June.
Jonathan, s. Robert Deborax, 2 June.
Ales, d. Richard Batte, 16 June.
Ann, d. John Start, 24 June.
Robert, s. Thomas Mathew, 29 June.
Clemens, s. Clemens Borham, 20 June.
Richard, s. "Mr. Richard King, parson of this towne," 27 June.
ffrancis, s. John Brown webster, 23 July.
Dorathy, d. Richard Edward, 30 September.
Sara, d. Thomas Harvy, 7 October.
Robert, s. John Perry, 12 Octobe.
Ann, d. Alexander Bulloyne, 12 October.
John, s. Robert Hogge, 2 November.
Thomas, s. Thomas Browne, 5 December.
Alexander, s. Thomas Plumbe, 21 December.
Mary, d. John Smith, 17 Januarie.
Sara, d. Robert Edward, 20 Januarie.
William, s. William Waford, 5 February.
Alice, d. John Ridgewell, February.
Robert, s. Hercules Newman, 1 March.
Margaret, d. Hercules Evans, 20 March.
1607 William, s. James Sheade, 6 Aprill.
Richard, s. Richard Payne, 22 Aprill.

Sara, d. John Langton, 4 July.
Thomas, s. Henry Laver, 18 Julie.
Elizabeth, d. Richard Titterill, 2 August.
Mary, d. John Brande, 15 Sept.
Ann, d. William Punnell,.31 Jan.
Thomas, s. Clemens Borham, 18 Feb.
Mary, d. Mr. Richard Kinge, parson, 21 Feb.
Dorothy, d. William Butcher, 28 Feb.
An, d. Thomas Gardiner, 3 March.
1608 John, s. Thomas Plumme, 20 June.
Henry, s. John Gips, 26 June.
Elizabeth, d. Jeremye Payman, 2 July.
Margaret, d. Robert Edwardes, Fullers, 28 July.
Margaret, d. William Waford, 10 August.
[17] Nathaniel, s. John Starte, 16 August.
John, s. Christopher Reade, 4 October.
Alice, d. William Bateman, 11 October.
John, s. Richarde Raven, 9 Nov.
Barbara, d. William Battye, 20 Dec.
Simon, s. John Foxe, 17 Jan.
Thomas, s. Thomas Gardiner, 11 Feb.
Francis, d. John Perry, 26 Feb.
1609 Nathaniel, s. Briget Brian, begotten in fornication, 17 April.
Thomas, s. Robert Devorax, 19 Aprill.
Richard, s. Hercules Evins, 14 May.
John, s. Jerome Perman, 10 October.
Michael, s. Michael Richardson, 19 October.
John, s. Mr. John Cracherood, 24 October.
William, s. Robert Pollard, 28 October.
James, s. James Shedd, 24 October.
William, s. Robert George, 11 January.
Thomas, s. Thomas Plumbe, of Olivers, 12 March.
Edward, s. John Start, 23 March.
1610 Alice, d. Robert Clerke, 29 March.
Elizabeth, d. Adlan Newman, 1 Aprill.
Richard, s. Richard Raven, 1 April.
William, s. Andrew Edward, 7 April.
Rose, d. John Drury, 7 April.
William, s. John Redgwell, 7 April.

Thomas, s. Robert Edward, of the fermen, 8 July.
Dorothy, d. Thomas Browne, 7 July.
Joane, d. Richard ffinch, 22 July.
[18] Judeth, d. William Horne, 10 August.
Margaret, d. William Dunnell, 12 August.
John, s. Richard Titterill, 12 August.
Ann, d. Robert Edward, of fullers, 22 August.
Ann, d. Clement Boram, 9 Sept.
Dorothy, d. William Bateman, 13 September.
Ann, d. Daniell Butcher, 11 November.
Thomas, s. John Perry, 2 December.
ffrancis, d. John Smith, 4 December.
Ann, d. Richard Butcher, 20 December.
Ellin, d. Robert Pollard, 13 January.
1611 John, s. John Hawksbee, 24 Aprill.
Johanna, d. Robert Denovan, 13 May.
Catheren, d. Robert Edward, jun., 21 May.
William, s. Atlas Evans, 23 May.
Thomas, s. William Battie, 30 May.
Daniell, s. Michael Richardson, 27 June.
William, s. Thomas Gardiner, 13 July.
Elizabeth, d. Thomas Plumb, 29 August.
Margaret, d. Richard Kinge, Doctor in Divinity, 19 September.
Ann, d. Thomas Browne, 3 October.
Thomas, s. Thomas Mathew, 18 November.
Marye, d. Christopher Roote, 21 January.
Richard, s. John Start, 28 January.
Margaret, d. John Quie, 25 February.
Elizabeth, d. Richard Edward, 27 February, 1616.
Sara, d. Henry Pettit, jun., 28 February.
Mary, d. William Overed, 18 March.
Sara, d. Robert Hogge, 23 March.
1612 Margaret, d. James Shedd, 4 Aprill.
Elizabeth, d. Robert Harrington, 14 Aprill.
Alice, d. Edward Clay, 11 May.
Elizabeth, ———, ———.
Margaret, d. Laurence More, 12 June.
John, s. Thomas Balie, 2 Julie.
William, s. Clement Boreham, 28 Julie.

Elizabeth, d. Jerome Perman, 30 Julie.
Dorcas, d. John Cracherood, gentleman, 6 August.
Georg, s. William Waford, 3 Sept.
William, s. William Bateman, 22 Sept.
Mary, d. Philip Ansell, 23 Sept.
Joseph, s. Michael Richardson, 14 October.
Elizabeth, d. Henrie Evans, 14 October.
Sara, d. Robert Pollard, 19 November.
Elizabeth, d. Thomas Gipp, 19 October.
Marie, d. Samuell Bateman, 13 January.
William, s. Richard Raven, 21 January.
John, s. Robt. Edwards, 7 Febr.
William, s. John Perry.
1613 Eliza, d. William Sparke, 25 March.
Susan, d. William Batty, 27 April.
Henery, s. William Smith, 23 May.
Robt., s. John Browne, 30 May.
Samuel, s. Andrew Edwards, 20 June.
Joanna, d. John Start, 28 July.
Annie, d. —— Medcalfe, 2 Aug.
[18] Mary, d. Robert Edwards, jun., [date illegible].
Sept. 30. Benjamin, s. Thomas Bateman.
October 3. Hellen, d. Samuel Dod.
November 5. Susan, d. William Overhead.
Decemb. 12. William, s. Edward Brown.
Jan. 6. Mary, d. Richard Kendhel.
Jan. 13. Samuell, s. Samuell Bateman.
Feb. 19. William, s. Robert Edwards the greater.
March 17. Susan, d. Mr. Richard Kinge, Dr. of Divinity.
1614 April 10. Susan, d. William Levit.
April 21. Mary, d. William Bacon.
April 26. Mary, d. Nathaniell Bateman.
June 2. Susan, d. Richard Harrington.
July 17. Laurance, s. Laurance Moore.
Sept. 18. William, s. Richard Butcher.
Sept. 1. Robert, s. Thomas Buttall.
Sept. 27. Mathew, s. William Butcher.
Octob. 2. John, s. Michaell Richardson.
Oct. 28. Josias, s. Josias Pollard.

Oct. 28. Alice, d. Robert Pollard.
Nov. 18. John, s. Nichelaus Evens.
Nov. 22. Jonathan, s. Christopher Roote.
Jan. 7. Edward, the base sonne of Elizabeth Grene.
Jan. 12. John, s. John Quie.
Jan. 19. Daniel, s. Richard Raven.
Jan. 31. Alice, d. William Sparke.
Feb. 2. Anne, d. Edward Clay.
Feb. 12. Stephen, s. Clement Boreham.
Feb. 21. John, s. William Edwards.
March 11. Ede, base daughter Adwy Fisher.
March 18. Thomas, s. William Bateman.
March 19. Daniell, s. Henry Bayley.
1615 May 4. Anne, d. William Smith.
May 2. William, s. Robert Medcalfe.
June 6. Hennery, s. Hennery Pettitt.
June 11. Ellen, d. Jeames Shedd.
Julie 21. John, s. John Dod.
June 27. Elizabeth, d. Robert Warner, jun.
Sep. 24. Robert, s. Thomas Butcher.
Nov. 2. Elizabeth, d. John Start.
Nov. 19. Nicholas, s. Nicholas Smith.
Decem. 2. Mary, d. John Perry.
Jan. 11. Barbery, d. John Pollard.
Jan. 21. Alice, d. Edward Moore.
Jan. 21. Daniel, s. Hennery Smith.
Jan. 28. Nathan, s. Nathan Bateman.
Jan. 30. William, s. William Cooper.
Decem. 21. Thomas, s. George Hogg.
Feb. 1. Sarah, d. Josias Pollard.
March 1. Martha, d. Jeremie Perman.
March 12. Jeames, s. Richard Kendall.
March 14. Elizabeth, d. Richard Harrington.
[19] 1616
April 2. Anne, d. William Sparke.
Rebecca, d. Daniel Dod.
May 21. Elizabeth, d. Robert Edwardes, jun.
June 2. Robert, s. Richard Titterill.
June 16. Rose, d. William Cooke.
June 24. Judith, d. Richard King.

Aug. 4. Susan, d. Thomas Gardner.
Aug. 13. Ellenor, d. Mr. John Cracherood.
Aug. 25. William, s. William Read.
Aug. 27. Robertah, d. Christopher Roote.
Sept. 1. Margaret, d. William Horne.
Novem. 24. Thomas, s. Thomas Brewer.
Decem. 1. Elizabeth, d. Giles Elsing.
Decem. 8. Robert, s. Robert Pollard.
Jan. 12. John, s. Robert Medcalf.
Jan. 19. Anne, d. John Drury.
Jan. 19. Susanna, d. Richard Gipps.
Jan. 23. William, s. William Bacon.
Feb. 26. Anne, d. Arthur Winterfloud.
Feb. 28. Elizabeth, d. Mihil[?] Osborne.
March 2. John, s. Richard Butcher.
March 4. Margaret, d. William Butcher.
March 19. William, s. William Levite.
March 19. Thomas & Alse, s. & d. Richard Raven.
1617 March 27. Susan, d. Henry Harrington.
April 8. Mary, d. Milvill Richardson.
April 21. Barbara, d. Robert Warner.
April 21. Lettice, d. John Start.
April 22. Edmund, s. Jeremie Parmeter.
May 22. Susauna, d. William Bateman.
June 15. Margaret, d. Thomas Robinson.
June 19. John, s. Thomas Dwe[?] alias Mathew.
Oct. 3. John Smithson, s. John Smithson.
Oct. 19. William, s. Edward Clay.
Oct. 23. Ric[d] Wight, s. Thomas Wight.
Nov. 9. Nicholas, s. —— Traylor.
Nov. 29. Emma, d. James Qui.
Dec. 4. Anne, d. William Edwards, Bradfields.
Dec. 10. William, s. Moyses Wallis.
Jan. 16. William, s. Henry Bayley.
Jan. 22. Elizabeth, d. Nathan Barman.
Jan. 25. Henry, s. John Dod.
Feb. 16. William, s. Daniell Dod.
March 1. Dennes, s. Giles Elsing.
1618 April 6. Sara, d. William Bard.
April 8. Susan, d. Thomas Trapner.

Mai 4. Susan, d. Richard Kendall.
[20] Mai 27. Henry, s. Samuel Smith.
June 15. Richard, s. Samuel Simons, gent.
June 28. Thomas, s. Richard Rane.
June 29. Joseph, s. Joseph Mariner.
June 30. John, s. Nicholas Smith.
July 2. William, s. William Smith.
Aug. 2. Mary, d. James Shed.
Sept. 17. Robert, s. Robert Edwards.
Sept. 29. ffelix, s. ffelix Torinne.
Oct. 1. Mary, d. John Pollard.
Oct. 4. William, s. Thomas Brewer.
Oct. 8. Mary, d. George Hogg.
Nov. 1. Joseph, s. John Simpson.
Nov. 10. Henry, s. Henry Petitt.
Nov. 19. Anne, d. Thomas Andrews.
Nov. 27. Thomas, s. John Start.
Jan. 6. Grisill, d. William Levett.
Jan. 14. Elizabeth, d. Jeremie Parmeter.
Jan. 27. Daniel, s. Michael Richardsonne.
Jan. 31. Henrie, s. Henrie Baylie.
Feb. 7. Henrie, s. William Read.
Feb. 9. Margarett, d. Robert Warner.
Feb. 19. Grace, d. William Cooke.
1619 March 29. Avis, d. William ffitch and Crissie his wife was baptized.
April 20. John, s. Robert Pollard & Ellen.
April 8. Richard, s. Thomas Wight & Edith.
April 17. Dorcas, d. Moses Wallis & Elizabeth.
April 25. Margaret, d. Clement Borham & Luce.
April 27. ffrances, d. Samuell Edwards, jun., & ffrances.
May 20. Elizabeth, d. Richard King & Margarett.
June 10. Mary, d. William Sparke & Catherine.
June 13. John, s. Michell Osburne & Marie.
July 7. William, s. Thomas Mathew & Margerie.
July 15. Anne, d. Laurence More & Elizabeth.
Aug. 15. Alice, d. Richard Butcher.
Oct. 3. Robert, s. Nathaniell Horne & Judith.
Novem. 9. Dorothy, d. Samuel Simons & Dorothe.

Novem. 18. Sara, d. William Bateman & ffrances [?]
Oct. 9. Anne, d. William Butcher, of Gainsford, & Margaret.
Decem. 17. Marie, d. Roger Hoyden & Sara.
[21] Decem. 20. Marie, d. Ambrose Tompson & Tomasyn.
Jan. 6. Joane, d. John Drury & Ester.
Jan. 13. Thomas, s. Josias Pollard & Sara.
March 7. Margaret, d. Thomas Trapnell & Anne.
1620 April 2. Robert, s. William Read & Alice.
April 8. Samuell, s. Samuell Hammond and Anne.
April 19. Thomas, s. John Allston & Anne.
April 20. Henry, s. Nathan Bateman & Mary.
April 20. George, s. Joseph Mariner & Mary.
April 26. Martha, d. William Edwards, de Bradfields, & Ane.
May 9. Elizabeth, d. Robert Trilne & Amis.
May 15. Frances, d. Clement Boreham & Luce.
May 22. Richard, s. Richard Edwards.
May 25. William, s. William Briant & Elizabeth.
June 3. Sarah, d. Richard Larke & Esther.
June 6. John, s. William Smith & Anne.
June 29. John, s. John Pollard & Elizabeth.
July 2. Thomas, s. Thomas Robinson & Margaret.
July 2. Anne, d. John ffisher & Bridgett.
July 6. John, s. George Hogge & Elizabeth.
July 6. Nathan, s. Richard Rane & Alse.
Aug. 5. William, s. William Sparke & Catherine.
Aug. 20. Henry, s. Robert Laver & Mary.
Sept. 19. Margaret, d. Edward Clay & Anne.
Sept. 19. Tomazine, d. ffelix Torrine & Mary.
Nov. 21. Elizabeth, d. Samuel Edwards & ffrances.
Dec. 17. Margaret, d. William Bacon & Marye.
Dec. 18. William, s. John Quie and Sara.
Jan. 5. Tomazin, base daughter of John Clarke & Elizabeth Browne.
Feb. 1. Esthamoth, s. Milvill Richards.
Feb. 10. John, s. Thomas Gardiner & Sara.
March 18. Elizabeth, d. Henry Petit & Sara.
[22]
1621 April 5. Elizabeth, d. Robert Warner & Elizabeth.

April 29. Jane, d. Samuel Simons & Dorothee.
William, s. William ffitche & Grizell, 17 June.
Martha, d. Robert Edwards, the younger, & Catherin, 21 June.
Thomas, s. Thomas Hawkins & Anne, 1 July.
Susan, d. Ralph Sewell & Mary, 29 July.
Susan, d. James Shed & Mary, 29 July.
Richard, s. Samuell Hamond & Anne, 9 August.
Margaret, d. John Simpson & Elizabeth, 17 Sept.
William, s. Richard Edwards & Mary, 10 October.
Mary, d. Jerome Parmenter & Ann, 19 August.
Peter, s. Robert Pollard & Ellenor, 27 October.
Joseph, s. William Butcher & Margarett, 15 November.
Thomasin, d. Henry Bayly & Judith, 2 December.
Thomas, s. Thomas Trapnell & Anne, 10 Dec.
Joseph, s. Mary Tilbroke, 20 January.
Marie, d. William Pamplin & Sarah, 3 March.
John, s. Roger Hoyden & Sarah, 5 March.
Allice, d. Giles Elsing & Elizabeth, 24 March.

[23]
1622 John, s. John Start & Allice, 26 March.
Thomas, s. William Bryant & Elizabeth, 31 March.
Robert, s. John Pollard & Elizabeth, 31 March.
Dorcas, d. Thomas Cratcherode, gent., & Susan, 18 Aprill.
Anne, d. Samuel Simons, gent., & Dorothy, 25 April.
Sarah, d. John ffisher and Brigett, 28 April.
Thomasin, d. Jeremy Pearmaine & Thomasin, 5 May.
Robert, s. William Levett & Susan, 20 June.
Samuel, s. John Dod & Mary, 8 July.
Elizabeth, d. Richard Kendall & Anna, 5 Sept.
Thomas, s. Henry Paine & Mary, 15 Sept.
William, s. Joseph Marriner & Mary, 18 Sept.
William, s. Thomas Buttoll & Elizabeth, 22 Sept.
John, s. John Drury & Hester, 5 December.
John, s. William Read & Allice, 15 December.
Samuel, s. Nathaniel Horne & Joane, 12 January.
Robert, s. Robt. Traylor & Avis, 16 January.
[24] Margarett, d. Samuell Hamond & Anne, 23 January.
Ellen, d. Clement Borum & Luce, 10 ffebruary.

Samuell, s. Richard Edwards & Mary, 9 March.
1623 Susan, d. Michael Richardson & Elizabeth, 1 April.
Mary, d. John Read & Mary, 15 Aprill.
Joane, d. Rafe Sewell & Mary, 2 June.
Susan, d. Henry Pettitt & Sarah, 20 June.
Margarett, d. Laurence More & Elizabeth, 10 August.
John, s. Thomas Cratcherode & Susan, 13 August.
Robert, s. Robert Warner & Elizabeth, 14 August.
Tomasin, d. Theodore Cole & Tomasin, 26 August.
Mary, d. Robert Wentford & Anna, 28 August.
William, s. William Edwards & Anne, 9 October.
Samuell, s. Samuell Symons & Dorothy, 29 October.
Jane, d. William Pamplin & Sarah, 14 December.
Richard, s. Robert Pollard & Ellenor, 1 February.
Mary, d. Barnard Sibly & ffrancis, 10 ffebruary.
Arnold, s. Arnold Wade & Mary, 12 ffebruary.
[25] Sarah, d. Thomas Hale & Susan, 25 ffebruary.
1624 Anne, d. John Mising & Anne, 2 May.
John, s. Robert Edwards & Katherin, 6 May.
John, s. William ffitch & Grizell, 13 June.
Elizabeth, d. John Gardiner, gent., & Jane, 15 July.
William, s. John Simpson & Elizabeth, 25 July.
Thomas, s. John Simpson & Elizabeth, 25 July.
Susan, d. John Quy & Sarah, 1 August.
Samuell, s. William Bryant & Elizabeth, 29 August.
Thomas, s. William Bacon & Mary, 28 September.
ffrancis, d. Edward Clay & Anne, 28 October.
Margarett, d. Richard Kendall & Anne, 16 November.
Grace, d. John Pollard & Elizabeth, 28 November.
Elizabeth, d. John Drury & Hester, 28 November.
Susan, d. Thomas Paynell & Grace, 2 Dec.
Henry, s. Henry Paine & Mary, 12 December.
Mary, d. Thomas Robinson, alias Butcher, & Margarett, 12 December.
Anne, d. Thomas Trapnell & Anne, 19 December.
Thomas, s. Richard Edwards & Mary, 19 December.
Elizabeth, d. Samuell Symons & Dorothy, 22 December.
[26] Susan, d. John Read & Mary, 3 January.
William, s. Roger Hoyden & Sarah, 11 January.

Theodore, s. Theodore Cole & Tamesin, 31 January.
Mary, d. Robert Warner & Elizabeth, 17 ffebruary.
Anne, d. Samuell Hamond & Anne, 17 March.
Susan, d. Richard Gyps & Susan, 20 March.
William, s. William Spark & Katherin, 21 March.
1625 Anne, d. Henrie Pettitt & Sarah, 28 Aprill.
Sarah, d. William Hart & Ellenor, 12 May.
Mary, d. Rafe Sewell & Mary, 17 May.
Agnes, d. Robert Wankford & Agnes, 4 July.
Elizabeth, d. John Warner & Elizabeth, 7 July.
George, s. George Gyps & Rebeckah, 10 July.
Thomas, s. Thomas Warner, the younger, & Dorothy, 28 July.
Sarah, d. William Read & Allice, 19 August.
Mary, d. Thomas Greene & Mary, 8 Sept.
Susan, d. Richard Lark & Hester, 3 January.
Samuell, s. Samuell Symons & Dorothy, 3 January.
Mary, d. Mary Edwards, 3 January.
Thomas, s. William Butcher & Margarett, 5 January.
Alice, d. William Batty & Elizabeth, 15 January.
George, s. Arnold Wade & Mary, 10 ffebruary.
[27] Mary, d. Robert Trailane & Anne, 21 ffebruary.
Jane, d. John Gardiner, gent., & Jane, 2 March.
John, s. William Edwards & Anne, 15 March.
Thomas, s. William Levett & Susan, 16 March.
Judith, d. Henry Baily & Judith, 19 March.
1626 William, s. William Wright & Winifride, 27 March.
William, s. John Start & Jael[?], 25 April.
Thomas, s. Robert Pollard & Ellenor, 14 May.
Susan, d. Thomas Cracherode, gent., & Susan, 31 May.
Thomas, s. John Dod & Marie, 9 July.
William, s. William ffitch & Grisell, 27 July.
Edward, s. William Bigge & Meliora, 7 September.
John, s. William Cook & Grace, 24 September.
Susan, d. Daniell Newman & Christian, 10 October.
William, s. William Cason & Marie, 24 October.
Samuell, s. Robert Edwards & Katherine, 24 October.
Susan, d. John Mising & Anne, 30 October.
Ralfe, s. Ralfe Sewell & Marie, 14 January.
Nicholas, s. Thomas Piner & Marie, 16 January.

Anne, d. John Bryant & Dorcas, 4 February.
Thomas, s. John Pollard & Elizabeth, 4 ffebruary.
William, s. Thomas Trapnell & Anne, 15 ffebruary.
[28] Anna, d. Theodore Cole & Tomasin, 24 ffebruary.
John, s. John Read & Marie, 4 March.
Anne, d. Edward Wade & Margarett, 4 March.
John, s. John Purchas & Elizabeth, 10 March.
1627 George, s. George Gyps & Rebeckah, 27 March.
Harlakinden, s. Samuell Simons, gent., & Dorothy, 7 Aprill.
Thomas, s. John Gardiner, gent., & Jane, 1 May.
Mathias, s. Robert Pollard & Ellenor, 29 Julie.
John, s. Thomas Emsden & Elizabeth, 9 Sept.
Rose, d. William Pamplin & Sarah, 16 Sept.
Priscilla, d. Roger Hoiden & Sarah, 23 Sept.
Elizabeth, d. Thomas Harvie & Susan, 30 Sept.
Thomas, ye base sonne of Marie Butcher, 21 October.
Samuel, s. Samuell Bell & Elizabeth, 18 November.
Thomas, s. William Sparke & Katherine, 20 November.
Lidia, d. Samuell Hamond & Anne, 13 November.
Robert, s. John Warner & Elizabeth, 24 Januarie.
Edward, s. Edward Clay & Anne, 10 ffebruarie.
John, s. Richard Larke & Hester, 17 ffebruarie.
Giles, s. Giles Elsing & Elizabeth, 17 ffebruarie.
Thomas, s. Henry Pettit & Sarah, 21 ffebruarie.
Marie, d. Richard Edwardes & Marie, 25 ffebruarie.
[29] John, s. Arnold Wade & Marie, 25 March.
1628 Elizabeth, d. John Purchas & Elizabeth, 27 March.
Susan, d. William Smith & Anne, 1 Aprill.
Richard, s. Thomas Cracherode, gent., & Susan, 22 Aprill.
Daniell, s. John Dod & Marie, 18 May.
Bridgid, d. John Gardiner, gent., and Jane, 27 May.
John, s. John Drurie & Hester, 22 June.
Elizabeth, d. William Edwards & Anne, 10 Julie.
Robert, s. Ralfe Sewell and Marie, 17 Julie.
John, s. Samuell Simons, gent., & Dorothy, 18 Julie.
Mathias, s. Mathias Gurton & Dorothy, 27 Julie.
John, s. John Gyps, junior, & Susan, 3 August.
Marie, d. Robert Reeid & Joan, 17 August.

Thomas, s. William ffitch & Grisell, 7 September.
Jael, d. John Start & Jael, 14 September.
Martha, d. Robert Warner & Elizabeth, 9 October.
John, s. John Laver & Marie, 6 November.
Elizabeth, d. Thomas Trapnell & Anne, 6 November.
John, s. Theodore Cole & Tamesin, 18 November.
Marie, d. Ellenor Hart, widdow, 2 Januarie.
Daniell, s. John Busie & Anne, 18 Januarie.
Marie, d. William Cooke & Grace, 15 ffebruarie.
Dorcas, d. John Bryant & Dorcas, 1 March.

1629
[30] John, s. George Gyps & Rebeckah, 7 Aprill.
ffrancis, d. Richard Edwards & Marie, 26 Aprill.
Jane, d. Robert Trayland & Avis, 1 May.
Samuell, s. John Read & Marie, 5 Julie.
Richard, s. Henrie Bailie & Jjudith, 19 Julie. [gust.
Robert, s. Samuell Symons, gent., & Dorothy, 7 Au-
Susan, d. John Gardiner, gent., & Jane, 18 August.
Thomas, s. Thomas Harvy & Susan, 1 October.
William, s. Elizabeth Browne, 4 October.
Susan, d. Samuell Bell & Elizabeth, 28 October.
William, s. Edward Tailer & Elizabeth, 22 November.
Dorcas, d. Robert Wentford & Marie, 1 December.
Thomas, s. Robert Pollard & Ellen, 6 December.
Marie, d. John Laver & Marie, 16 December.
William, s. William Murkin & ffrancis, 21 January.
Sarah, d. William Pamplin & Sarah, 24 Januarie.
Susan, d. Arnold Wade & Marie, 26 Januarie.
Martha, d. Henrie ffrench & Martha, 3 March.
Susan, d. John Bryant & Susan, 7 March.

1630
[31] Susan, d. John Purchas & Elizabeth, 30 March.
Daniell, s. John Start & Jael, 15 Aprill.
John, s. John Easterford & Emme, 27 May.
George, s. Thomas Trapnell & Anne, 6 June.
Anne, d. John Busie & Anne, 20 June.
John, s. John Newman & Sarah, 21 September.
Marie, d. John Gardiner, gent., & Jane, 27 October.
Sarah, d. Steven Warner & Sarah, 7 November.
William, s. Henrie Pettit & Sarah, 11 November.

Daniell, s. Barnard Sibly & ffrancis, 14 November.
Marie, d. John Anderson & Elizabeth, 28 November.
Susan, d. John Gyps, jun., & Susan, 28 November.
ffrancis, d. Richard Whiffin & ffrancis, 5 December.
Hester, d. George Gyps & Rebeckah, 12 December.
William, s. Thomas Emsden & Elizabeth, 6 Januarie.
Samuell, s. William Edwards & Anne, 24 ffebruarie.
Robert, s. William Edwards, junior, of ffullers, & Martha, 13 March.
1631 ffrancis, s. John Simpson & Elizabeth, 12 Aprill.
William, s. John Warner & Elizabeth, 21 Aprill.
[32] Anne, d. John Hale & Margarett, 24 Aprill.
Marie, d. John Bryant & Dorcas, 28 Aprill.
Hester, d. Robert Warner & Elizabeth, 12 May.
Anne, d, Robert Edwards & Luce, 12 May.
Susan, d. Thomas Pollard & Susan, 22 May.
Robert, s. Robert Wentford & Marie, 12 June.
Elizabeth, d. Theodore Cole & Tamesin, 19 June.
John, s. John ffitches, the younger, & Elizabeth, 17 Julie.
Elizabeth, d. Thomas Harvy & Susan, 31 Julie.
Susan, d. John Bellowes & Susan, 11 August.
ffrancis, d. John Busie & Anne, 29 September.
Martha, d. John Read & Maria, 16 October.
Hester, d. Richard Larke & Hester, 28 October.
Henrie, s. Henrie ffrench & Martha, 10 November.
Grace, d. John Gardiner, gent., & Jane, 16 November.
Marie, d. John Laver & Marie, 29 November.
John, s. John Purchas & Elizabeth, 19 ffebruary.
Elizabeth, d. Edward Tailer & Elizabeth, 11 March.
Lydia, d. Arnold Wade & Marie, 13 March.
[33]
1632 Susan, d. John Easterford & Emme, 27 March.
Anne, d. Thomas Trapnell & Anne, 10 May.
Elizabeth, d. Thomas Emsden & Elizabeth, 13 May.
Thomas, s. Samuel Bell & Elizabeth, 27 May.
Elizabeth, d. Edmund Drury, gent., & Elizabeth, 13 June.
William, s. Samuell Symons, gent., & Dorothy, 22 June
William, s. William Pamlin & Sarah, 22 July.

Samuell, s. William Cooke & Grace, 1 August.
Marie, d. John Bellowes & Susan, 16 September.
Elizabeth, d. William Butcher, junior, & Anne, 2 October.
Henry, s. William Cason & Marie, 29 May.
Dorothy, d. John Perrie, the younger, & Katherin, 7 October.
Elizabeth, d. John Anderson & Elizabeth, 28 October.
Rose, d. John Busie & Anne, 18 November.
Margaret, d. John ffisher & Allice, 18 November.
John, s. Thomas Greene & Elizabeth, 6 December.
Samuel, s. Robert Wentford & Marie, 18 December.
[34] ffrances, s. Robert Trayland & Avis, 23 December.
Thomas, s. William Redgwell & Jone, 10 ffebruary.
Richard, s. Samuell Smith & Joice, 12 ffebruary.
Katharine, d. William Chadderton & Katharine, 17 ffebruary.
Anne, d. ffrancis Martin & Marie, 19 ffebruary.
ffrancis, d. William Murkin & ffrancis, 28 ffebruary.
Anne, d. Robert Dawson & Anne, 4 March.
Grace, d. Thomas Pollard & Susan, 24 March.
1633 Anne, d. John Gardiner, gent., & Jane, 8 Aprill.
Robert, s. Peter Hale & Susan, 23 Aprill.
William, s. William Browne & Dorothy, 26 May.
Thomas, s. Thomas Warner, the younger, & Agnes, 2 June.
Marie, d. John Gyps & Susan, 9 June.
John, s. John ffitches, the younger, & Elizabeth, 2 July
John, s. John Bryant & Dorcas, 14 July.
[35] Samuell, s. Richard Whiffin & ffrancis, 29 September.
Judith, d. Barnard Sibly & ffrancis, 29 September.
William, s. Mathew Hart & Christian, 6 October.
Marie, d. Thomas Harvy & Susan, 13 October.
James, s. James Windle & Anne, 5 November.
John, s. Thomas Sibly & Agnes, 10 November.
Sarah, d. John Start & Jael, 10 November.
Martha, d. Robert Edwards & Luce, 17 November.
Roger, s. Samuell Symons, gent., & Dorothy, 5 December.
Dorcas, d. John Laver & Marie, 12 December.

Katharin, d. Clemens Boreham, junior, & Katharin, 21 Januarie.
Richard, s. Richard Titterell, the younger, & Jone, 2 ffebruary.
Marie, d. Theodore Cole & Tamesine, 16 ffebruary.
Robert, s. Josias Pollard & Anne, 15 December.
Dorcas, d. John Read & Marie, 18 March.

[36]

1634 Margarett, d. Allice Clay, 25 March.
William, s. Thomas ffowle, gent., & Barbara, 4 May.
John, s. Richard Edwards & Marie, 3 May.
Anna, d. Anna Cresling, 20 July.
Susan, d. William Edwards & Anne, 7 August.
John, s. John Bordman & Rose, 7 September.
Sarah, d. John Easterford & Emme, 10 October.
William, s. John Perry, the younger, & Katharin, 20 October.
Susan, d. Ralfe Turner & Susan, 30 November.
Elizabeth, d. Thomas Greene & Elizabeth, 30 November.
Marie, d. Henry ffrench & Martha, 21 December.
Marie, d. John ffisher & Allice, 28 December.
Anne, d. John Dollar & Anne, 28 December.
William, s. John Bryant & Dorcas, 29 January.
Marie, d. William Warner & Marie, 29 January.
[37] Henrie, s. John Busie & Anne, 2 ffebruary.
John, s. William Pamplin & Sara, 8 ffebruary.
Marie, d. ffrancis Martin & Mary, 5 March.
1635 Thomas, s. Thomas Pollard & Susan, 30 Aprill.
Samuell, s. Thomas Emsden & Elizabeth, 14 May.
Richard, s. Thomas ffowle, gent., & Barbara, 2 June.
Martha, d. William Browne & Dorothy, 30 June.
Marie, d. Nathanaell Paul & Marie, 19 Julie.
Lydia, d. William Edwards, of ffullers, & Lydia, 18 August.
Tabitha, d. John Anderson & Elizabeth, 23 August.
Anne, d. William Redgewell & Jone, 23 August.
Jonas, s. James Windle & Anne, 17 September.
John, s. John Bellowes & Susan, 27 September.
Anne, d. Thomas Warner & Agnes, 4 October.

[38] Elizabeth, d. Nathanael Thurston, gent., & Marie, 9 October.
John, s. William Murkin & ffrancis, 11 October.
Thomas, s. Thomas Roote & Anna, 10 November.
Anna, d. Edward Tailer & Elizabeth, 22 November.
William, s. Robert Edwards & Luce, 29 December.
Susan, d. Thomas Harvy & Susan, 13 December.
Steven, s. Steven Warner & Sara, 27 December.
ffrancis, d. Thomas Edwards & Elizabeth, 21 ffebruary
ffrancis, reputed son of Thomas Smith & Margaret his wyfe, 13 March.
William, s. Henrie Smith & Sara, 15 March.
Marie, d. Thomas Trapnell & Anne, 15 March.
ffrancis Gall is my name and eeigth son of ——*
1636 John, s. Samuell Bateman & Elizabeth, 10 Aprill.
Clemens, s. Clemens Borham, junior, & Katharine, 19 Aprill.
Richard, s. John Gyps & Susan, 5 May.
Marie, d. William Greene & Marie, 15 May. [May.
[39] Olive, d. John Perrie, the younger, & Katharine, 29
Mathew, s. John ffitches & Elizabeth, 29 June.
John, s. John Titterell & Dorcas, 3 July.
Marie, d. Robert Wentford & Marie, 14 August.
Sara, d. Thomas Wood & Sara, 4 September.
Elizabeth, d. Richard Titterill & Jone, 25 September.
Robert, s. Robert Maltiward, gent., & Elizabeth, 1 October.
Ann, d. William Cooke & Grace, 15 October.
Ann, d. Richard Whiffer & ffrancys, 23 October.
Ann, d. William Mere & Sara, 28 October.
Richard, s. Josias Pollard & Anne, 1 November.
William, s. Thomas Greene & Elizabeth, 15 December
Thomas, s. George Earles & Elizabeth, 1 January.
John, s. John ffisher & Alice, 8 January.
Mary, d. William Browne & Dorothy, 18 January.
Rebecca, d. Elizabeth Sparke, base born, 4 February.
John, s. John Kinge & Margarett, 10 February.
Dorothy, d. ffrancis Martin & Mary, 5 March.
Elizabeth, d. Michael Richardson, jun., & Mary, 16 March.

---

*The rest of the entry is blotted and illegible.

1637 John, s. John Bryant & Dorcas, 11 April.
[40] Benjamin, s. John Reade & Mary, 25 April.
Susan, d. Thomas Mathews & Susan, 15 May.
Sara, d. Henry Smith & Sara, 30 May.
Robt., s. John Esterford & Emmey, 23 July.
Thomas, s. John Start & Jael, 27 August.
Robt., s. Robt. Edwards, jun., & *Luce, 17 Septemb.
William, s. John Bellowes & Susan, "the same day and year."
Elizabeth, d. John Busie & Anne, 29 Octob.
William, s. William Redgewell & Joane, 12 Novemb.
Thomas, s. Robt. Maltiward, gent., & Elizabeth, 26 Novemb.
Mary, d. Thomas Roote & Anna, 3 Decemb.
Robert, s. John Laver & Mary, 21 Decemb.
Mary, d. William Murkin & Frances, 17 Decemb.
William, s. William Borde, 28 Decemb.
John, s. Henry Ewins & Susan, 18 Janu.
Dorothy, d. Thomas Harvey & Susan, 21 Janu.
Samuell, s. Thomas Emsden, 12 Feb.
Robert, s. Clement Borham & Catherine, 4 March.
1638 Mary, d. Michael Richardson, jun., & Mary, 15 April.
Anne, d. William Smith & Margaret, the same day.
John, s. Josias Pollard & Anna, 29 June.
John, s. John Overed, Clerke & Christian, 3 July.
Samuel, s. Samuel Bateman, the younger, & Hannah, 29 August.
Thomas, s. John Gyps & Susan, 28 August.
John, s. Thomas ffowle, gent., & Margaret, 2 Septem.
William, s. William† & Margaret, 27 Septem.
Mary, d. William Alston, gent., & Mary, 27 Septem.
[41] Timothie, s. John Kinge & Margaret, the day aforesaid
William, s. William Greene & Mary, 15 Octob.
Anna, d. Richard Kempe & Mary, 11 Octob.
Mary, d. Henry Smith & Sarah, 5 Novemb.
Henry, s. Henry Laver & Mary, 15 Novemb.
William, s. John ffitches & Elizabeth, 13 Decemb.
Mary, d. William Borham & Margaret, 6 Janu.

---

*The name Elizabeth is erased.

†The name is much rubbed, the first letter is H, the end is plainly *iball*, and the whole name probably *Huddiball.*

Henery, s. Thomas Greene & Elizabeth, 13 Janu.
Jonathan, s. Thomas Roote & Anna, 2 Feb.
Martha, d. Laurence More & Martha, 17 Feb.
Mathew, s. Richard Edwards & Mary, 19 Feb.
———, –. Edward Taylor & Elizabeth, 25 Feb.
Elizabeth, d. Robt. Malteward, gent., & Elizabeth, 5 March.
1639 Elizabeth, d. Thomas Cornell & Katherin, 30 March.
Richard, s. Richard Cant & Elizabeth, 16 April.
Dorcas, d. John Titterill & Dorcas, 11 May.
Thomas, s. Martin Olley & Anne, 23 May.
———, d. Richard Cobb, 16 September.
Susan, d. Thomas Woode & Sarah, 20 September.
Samuel, s. Samuel Bateman, jun., 20 October.
Dorcas, d. Henery Laver, jun., & Mary, 27 Octob.
Mathias, baseborn child of Susan Shedd, 19 Novemb.
Elizabeth, d. John Overed, Clerk & Christian, 20 Novem
Mary, d. Oliver Keene & May, 26 December.
ffrances, d. John Briant & Dorcas, 14 January.
Hannah, d. Thomas Mathewes & Susan, the same day.
Samuel, s. John Weeber & ffrances, 26 January.
Moses, s. Moses Harrington, 23 Feb.
Margaret, d. John Kinge & Margaret, 23 Feb.
Grace, d. Gabriel Shedd & Agnes [?], 28 Feb.
[42] Martha, d. Michael Richardson, jun., & Ann, 8 March.
Daniel, s. Daniel Gurten, 8 March.
Mary, d. John Esterford & Emvira, the same day.
Dorothy, d. John ffisher & Alice, 10 March.
Henery, s. Henery Smith & Sarah, 12 March.
Anne, baseborn child of Anna Winterfloud, 23 March.
Martha, d. John Bellowes & Susanah, 4 August.
1640 William, s. William Redgewell & Joane, 4 April.
Robert, s. Robt. Right & Mary, 7 April.
———, d. Richard Titterill, 20 May.
John, s. —— Anderson & Elizabeth, 31 May.
Peter, s. Thomas Warner & Anne, 5 July.
Hester, d. James Chaplyn & Hester, 12 August.
Joseph, s. William Murkin & ffrances, 27 Septemb.
Thomas, s. Richard Kempe & Mary, 28 Octob.
Margaret, d. William Huddiball & Margaret, 15 Nov.

Mary, d. Robert Ailcoke & Elizabeth, 20 November.
Elizabeth, d. Laurence More & Martha, 29 Novemb.
William, s. Richard Whiffin & ffrances, 6 Demb. [*sic.*]
Robert, s. Richard Cant & Elizabeth, 31 January.
Thomas, s. Thomas Woode, 7 March.
Mary, d. William Bacon & Mary, 9 March.
1641 John, s. Dennis Elye & Sarah, 28 March.
Samuel, s. Samuel Bateman, the elder, & Elizabeth, April 27.
Thomas, s. Samuel Bateman, the younger, & Hannah, the same day.
Samuell, s. Robert Lynsell & Barbara, last day of May.
Jane, d. Robert Mathew & Margaret, 21 June.
Susanna, d. William Alstone, gent., & Mary, 25 Sept.
Anne, d. Marten Olley, 10 Octob.
Michael, reputed s. Mihill Brewster, baseborn of — Quy, 28 Octob.
Hanna, d. James Knidall & Hanna, 20 July.
[43] Anne, d. Edward Tayler, November.
Susanna, d. William Smith, December 6.
Elizabeth, d. John Kinge & Margaret, December 25.
John, s. John Weeke & Frances, Jan. 26.
Robert, s. Robert Edwards & Mary, Feb. 20.
Thomas, s. Richard Kempe, Oct. 10.
Elizabeth, d. Oliver Keene & Mary, 2 day Feb.
1642 William, s. William Raven, Feb. 20.
Mary, d. Jeremy Piper, Aprill 2.
Samuel, s. Samuel Bridge, April 20.
James, s. James Chaplyn, April 23.
Mary, d. Thomas Mathews, April 26.
Elizabeth, d. John Stone, April 28.
Joseph, s. William Warner & Mary, June 24.
Martha, d. Clement Boreham, & Katheren, 2 Octob.
Anne, d. Thomas Cornell & Katheran, 19 October.
William, s. William Alston, gent., & Mary, 5 Feb.
James, s. James Shed, 5 Feb.
Anne, d. Laurance Moor, 23 March.
1643 Mary, d. Thomas Emsden & Elizabeth, 10 May.
James, s. James Kendall & Johannah, 6 August.
Elizabeth, d. William Smith & Elizabeth, 20 October.

William, s. Robert Edwards & Mary, 24 October.
Rose, d. Nathan Bateman, 28 October.
Sarah, d. Thomas Cornell & Katheran, 25 July.
Abigail, d. Richard Kemp, 15 Decemb.
Jemima, d. William Alston, gent., & Mary, Feb. 12.
Elizabeth, d. John Esterford & Emery, Feb. 26.
ffrancis Gall.

1644 Susan, d. Martin Olly & Anne, June 7.
William, s. William Warner & Mary, June 18.
[44] James, s. James Chaplin & Esther, July 2.
Mary, d. Nathan Bateman & Rose, Sept. 12. [30.
Thomas, s. Mordant Cratchrood & Dorothy, Novemb.
Thomas, s. Thomas Edward & Hannah, Janu. 5.
John, s. Henry Laver, senior, & Mary, feb. 2.
Henry, s. Henry Gyps, feb. 18.
Hannah, d. John Bellows, Feb. 28.
1645 Elizabeth, d. William Raven, April 8.
Elizabeth, d. Edward Earle & Katherin, April 14.
Robert, s. Henry Laver, junior, & Susan, feb. 20.
Sarah, d. John Start, jun., and Avis, May 4.
Thomas, s. Thomas Green & Anne, 12 May.
Martha, d. William Smith & Margarett, June 4.
William, s. William Smith & Elizabeth, July 8.
Elizabeth, d. William Raven & Mary, Oct. 7.
William, s. William Mathew & Anne, Oct. 20.
Robert, s. Henry Eivens & Alice, Feb. 10.
[45] Mary, d. John Overed, minister, & Penelope, 24 May.
Cornelius, s. Oliver Reave & Mary, 28 May.
Mary, d. Robert Edwards & Mary, the same day.
ffrances, d. William Green & Mary, June 12.
Margaret, d. Robert Butcher, 2 July.
Laurance, s. Laurance Moor & Martha, 12 July.
Joanna, d. Robert Gurton, 27 July.
Martha, d. Richard Kemp, 27 December.
Thomas, s. William Huddeball, 16 January.
1646 Anne, d. Thomas Edwards & Anne, 6 April.
Jeremiah, s. Jeremiah Piper & Sarah, 20 June.
John, s. Clement Boreham & Katherin, 26 June.
Grace, d. Moses Harrington, 17 May.
Elizabeth, d. Thomas Borham, Octob.

Jonathan, s. John Overed, minister, & Penelope, 29 August.

Robert, s. Richard Titterill, 2 October.

Susan, d. Thomas Emsden & Elizabeth, June 24.

John, s. George Earle & Katherin, Sept. 29.

Joanna, d. John Alston, gent., Sept. 21. [tob.

Anthony, s. Mordant Cratchrood & Dorothy, 23 Oc-

John, s. John Start and Avis, Septemb. 6.

Thomas, s. Robert Butcher, Septemb. 15.

Thomas, s. Thomas Simson, 26 febr.

John, s. Richard Pepper, 5 March.

Robert, s. William Edwards & Elizabeth, 25 January.

John, s. Robert Edwards & Mary, 10 Novemb.

[46] Henry, s. Henry Laver, junior, & Susan, 20 March.

William, s. Richard Warner & Rachell, 18 March.

1647 Sarah, d. John Overed & Penelope, 30 June.

Robert, s. Michael Brewster, 2 August.

William, s. William Deeks, 12 August.

Robert, s. Robert Earle & Mary, Sept. 5.

Robert, s. Robert Pollard & Abigail, Sept. 20.

Sarah, d. Dennis Ely & Sarah, Dec. 27.

Robert, s. Christopher Erle, Esq.,* & Mary, February 19.

John, s. Robert Rust & Mary, February 27.

---

*This name is found elsewhere. Leland Duncan, Esq., F. S. A., writes: "I see that in one of the appendices to Dr. Shaw's 'History of the Church of England during the Commonwealth,' Toppesfield is spelt Topfield." He gives the name of the man who was intruded into the rectory and names of the "elders".——10 July, 1648, Mr. Overed to topesfield, Essex. Commons Journals, vol. v., p. 651; Lords Journals, x, p. 404.

An attempt was made to divide Essex into "Classes" for Presbyterian purposes. The 10th. Classis, called the Classis of Hincford, contained "Topfield," minister, Mr. Jo. Overed; Elders, Christopher Earle, Esq., Mr. Samuel Smith, and Robert Wentford. All these names are found in the register.

John Overed (erroneously called Thomas in the list of rectors exhibited in Toppesfield Church, and in copies made from that list) is first described in the registers as Clerke. He was a curate in charge of the parish during the time that Dr. Burnell was rector, the name of his wife at this time was Christian. She died (burials, Nov. 20, 1639), he afterwards married Penelope (May 24, 1643), where he is first described as minister. Probably he received the Rectory as a reward for his religious pliancy.

ffrancis Gall.
1648 ffrancis Gall.
John, s. Jeremiah Piper & Mary, June 20. [24.
Hannah, d. John Overed, minister, & Penelope, July
Daniel, s. Daniel Richardson & Rachel, May 10.
John, s. Thomas Green and Anne, June 28.
Thomas, s. Thomas Miller & Mary, July 4.
Thomas, s. Thomas Boreham & Elizabeth, July 5. [7.
Mary, d. Morduant Cracherood, gent., & Dorothy, July
Solomon, s. William Alston, gent., & Mary, July 18.
William, s. Samuel Bridge & Mercy, August 28.
John, s. Richard Titterell, Sept. 20.
Margaret, d. Thomas Matthew, Oct. 29.
[47] Mary, d. James Shed, January 2. [June.
1649 Penelope, d. John Overed, minister, & Penelope, 29
Matthew, s. Matthew Edwards & Ann, 12 May.
Elizabeth, d. John Pollard & Elizabeth, June 20.
Thomas, s. Thomas Green, June 21.
Philip, s. Thomas Winterflood & Mary, July 7.
Anne, d. John Smith & Margaret, Aug. 30.
Joseph, s. Richard Warner & Rachell, Feb. 5.
Daniel, s. Henry Laver & Susan, Janu. 1.
William, s. William Wright, Janu. 7. [27.
1650 Morduant, s. Morduant Cratchrood & Dorithy, March
Thomasin, d. William Edwards & Elizabeth, May 2.
Elizabeth, d. John Alston, gent., May 15.
Ann, d. Thomas Green, June 23.
Thomas, s. Moses Harrington, July 20.
Sarah, d. Edward Bointell & Sarah, Dec. 10.
Elizabeth, d. William Matthew & Ann, Dec. 15.
Samuel, s. Thomas Miller & Mary, Jan. 9.
John, s. John Stanes & Hester, the same day.
Samuel, s. Daniel Richardson & Rachell, Jan. 16.
John, s. Samuel Bridge & Mercy, Jan. 20.
Dorcas, reputed daughter of William Berd & baseborn of Dorcas Drury, July 10.
John, s. John Seaman & Rebecca, 12 July.
Samuel, s. Robert Earles & Mary, 6 August.
John, s. Thomas Simson & Margaret, feb. 18.

MARRIAGES.

---

[72]
1560 John Cirke & Ellin Buttall, 9 May.
William Pollard & Margaret Lende, 23 May.
Steven Titterell & Elizabeth Maysant, 27 May.
1561 Thomas ffytchs & Katherine Rayner, 7 December.
Michael Tongue & Alice Alwin, 3 October.
1562 Christopher ffiche & Joane Browne, 8 June.
Robert Edward & Marie Parker, 15 August.

The above seven entries are written (evidently copied) on the bottom of a page of baptisms; the next leaf has been cut out.

ffrancis Gall. Churchwarden.
ffor Yearre 1688.

[75]
1598 George Rule & Joane Wells, 30 April.
Richard Titterell & Anne Thetford, 22 June.
William Roger & Sarah Wast, 26 November.
1599 John Clarke of Radwinter, wid., & Margerie Berding single w., 11 April.
William Snelling & Margaret Chatwood, 10 August.
John Bayly, widd. & Mary Tilbish, widdowe, 29 August.
John Choate & Mary Grettit, 8 January.
Henry Smith & Mother Coozin, 10 December.
Robert Spiltimber & Thomasin Boultin, 19 October.
1600 David Playle & Margarett Lamson, 21 April.
Edmund Cant & Grace Welles, 8 May.
Henry Cirke & Rebecca Aylett, 28 October.
1601 Michael Clerk & Margaret Webb, 28 Octobe.

Robert Clerke & Joanne Trappe, 30 January.
1602 Samuell Edwards & Cicely Big, 7 November.
1603 Thomas Wilcocke & Ann Brown, 1 November.
Jeffry Harvy & Rebecca Cirk, 7 November.
James Weager & Lettice Bunting, 5 November.
John Smyth & Mary Edwards, 20 ffebruary.
1604 Josiah Collin & Elizabeth Amis, 20 April.
Robert Edwarde & Elizabeth Warner, 8 May.
Widdow[r]. Brown & Christian Pettit, 24 Julie.
John Cooke & Mary Horrold, 7 October.
1603 James Tayler & Triphena Borham, 23 August, 1603.
1604 Thomas Chatterton & Rose Perry, 16 January.
1605 Thomas Dawkins & Agnes Hart, 1 May.
Henry Snelhok & Goane Chapman, 7 May.
William Evans & Anne Wastell, 16 July.
Thomas Mathew, alias Micke, & Margery Boram, 4 August.
John Crawborrow & Ellen Evanfeill, widdow, 29 August.
Clement Borham & Lucy Crawborrow, 28 October.
John Pollard and Alice Edwards, 27 ffebruary.
1606 Henry French & Anne Perry, 13 July.
Thomas Horrold & Mary Cosin, 14 October.
John Brande & Mary Buttoll, Nov. 27.
John Godwin & Penelope Evered, Jan. 8.
1607 John Quye & Sara Shedd, April 18.
John Watson & Christian Reede, Sept. 14.
Thomas Clearkson & Mary Jolly, Sept. 28.
Jeremye Paxman & Elizabeth Reede, Oct. 8.
Richard Hamond & Joan Leony, Oct. 18.
1608 Henry Laver & Mathewe ———, June 30.
John Mayor & Elizabeth Everid, July 22.
Roger Ale & Margaret Browne, July 29.
George Lowimyr & Mary Bunting, Oct. 9.
Henry Petitte & Sara Edwards, Nov. 3.
[76] Thomas Javleton & Clement Bennet, Novemb. 30.
1609 John Wallaker & Margaret Warner, January 22.
William Overed & Mathew Bateman, January 23.
1610 John Gofeild & ffrancis Edwarde, 7 May.
John Bernard & Elizabeth Jollie, 2 Julye.
Robert Perry & Elizabeth Chamberlayne, 24 September.

Laurence More & Elizabeth Gippe, 13 October.
1611 Philip Anstell & Alse Edward, 27 May.
Thomas Gooddin & Margeret George, 1 July.
Samuel Bateman & Mary Berdin, 8 August.
1612 Thomas Onion & Debora Bateman, 21 April.
Nathan Bateman & Marie Livermore, 24 June.
Arthur Winterfloud & Clement Jurden, 3 August.
William Ward & Elizabeth Billiwood, alias Bust, 20 Sept.
1613 Edward Riche & Susan Warner, November 1.
John Clarke & Rose Laver, March 7.
1614 Gregory Warner & Elizabeth Edwards, May 8.
John Dod & Mary Amye, May 12.
John Ewer & Anne Lawer, May 29.
1615 Richard Gips & Susan Bateman, May 13.
Samuel Bland & Elizabeth Harvey, May 18.
Jeremie Pearman & Thomasin French, June 23.
William Cooke & Grace Cant, Julie 23.
Thomas Brewar & Susan Edwards, Jan. 9.
1616 Giles Elzing & Elizabeth Hide, April 25.
William Edwards & Anne Parke, May 27.
1617 Peter Hall & Susan Brian, Oct. 23.
John Simpson & Elizabeth ffarrar, June 18.
George Gurton & Alice Cirke, June 23.
Robert Frost & Judith Sibly, Oct. 28.
Ambrose Tompson & Thomasin Gips, November 9.
Thomas Andrews & Katherene Bragg, June 23.
1618 Thomas Bromley & Martha Webb, May 7.
Samuell Wike & Anne Sible, May 21.
Samuell Edwards & ffrances Greene, June 24.
Daniel Smith & Tomazin Smith, July 2.
William ffitch & Grisell Holton, Oct. 23.
[77] Thomas Ruggles & Elizabeth Barnard, Octob. 6.
Hercules Evans & Alice Cant, Nov. 5.
Henry Smith & Sarar Bigg, Nov. 23.
John Clay & Joane Rudland, wid., Jan 12.
Edward Bateman & Prudence Mosse, ff. 9.
1619 John Prackett & Thomazin Cordar, July 8.
July 31. Joseph Jackson & Margarett Read.
Sept. 27. Jhon ffisher & Bridgett Leman.

Sept. Richard Larke & Esther Cornhill.
Feb. 2. Raphe Sewell & Marie Butcher were buried[?].
1620 Sept. 29. Steven Waterworth & Alice Amie.
Nov. 28. John Start & Alice Dod, wid.
Jan. 28. Edward Mariott & Ester Bateman.
1621 April 19. William Browne & Anne Pettit.
John Most & Elizabeth Selling, 6 July.
Edward West & Alline Hostler, 2 August.
1622 John Read & Mary Warner, 10 June.
William Wright & Winifred Plaile, 24 June.
1623 Arnold Wade & Mary Bateman, 25 Aprill.
John Mising & Anne Cooke, 24 July.
William Butcher & Anne ffurmin, 2 Octobe. [ber.
Jeremie Pearmaine & Tamesin Thompson, 2 Novem-
1624 Jeremie Sansum & Lidia ffinch, 6 Aprill.
Elias Rayner & Barbara Tongue, 1 August.
Thomas Read & ffrances Tompson, 28 November.
1625 John Start & Elizabeth Baily, 5 ffebruary. [ruary.
[78] Robert Grant, alias Bryant & Elizabeth Hog, 21 ffeb-
1626 John Perrie & Rose Humfrey, Nov. 1.
1627 Joseph Jackson & Thomasin Chamberlaine, 2 August.
Mathias Gurton & Dorothie Hog, 5 November.
John Bellowes & Anne Start, 17 Januarie.
1628 Richard Wright & Dinah Smith, 4 August.
1629 Richard Lot & Rose Bun, 7 May.
Thomas Randoll & Barbara Davie, 21 Sept.
Robert Edwards & Luce Siblie, 29 Sept.
Robert Stevens & Ellen Hart, 5 October.
Henrie Gyps & Elizabeth Harrington, 24 November.
Richard Shellie & Elizabeth Start, 24 Januarie.
1630 John ffisher & Allice Smith, 8 July.
John Turner & Anna Cole, 12 Julie.
John Dollar & Anne Browne, 10 August.
John ffitches & Elizabeth Greene, 7 October.
1631 William Edwards, of ffullers, & Rachell Bridge, 15 November.
Henry Clarke & Clemens Winterfloud, 20 November.
John Onyon & Elizabeth Smith, 20 Januarie.
1632 Thomas Warner & Anne Greene, 11 June.

John Jaggard & Susan Roote, 3 January.

[79]

1633 John Bordman & Rose Drury, 22 July.
John Skelie & Dorothy Butcher, 25 July.
William Warner & Marie Edward, 15 August.
John King & Anne Kendall, 8 October. [May.
1634 Robert Mathew, alias Wix, & Margarett Boote, 27
William Greene & Marie Lithermore, 6 November.
William Board & Sarah Argent, 24 January.
1635 Thomas Birkner & Sara Wood, 30 Aprill.
William Clarke & ffrances Edward, 30 July.
Robert Maltiward & Elizabeth Cracherode, 26 August.
Henrie Holton & Margarett Evans, 29 September.
Nicholas Argent & Marie Titterell, 20 October.
William Perrie & Abigail Hancock, 28 October.
1636 Thomas Wood & Sara Shed, 28 November.
Henry Evens & Susan Ostler, 17 October.
John Parmane, the younger, & Elizabeth Butcher, 8 October.
1637 Thomas Bacon & Priscilla Badcocke, 20 April.
William Huddiball & Margaret Shed, last day of November.

[80]

1638 John Wicks & ffrances Perrye, 19 October.
George Sharp & Mary Drury, 23 October.
Nathaniel Kent & Anne Turner, 5 November.
Nathan ffrogg & Joane Griggs, 24 January.
1639 Robert Rust & Mary Bateman, 23 May.
Henry May & Elizabeth Harding, 24 June.
Aron Butcher & Elizabeth Edwards, 30 October.
1640 William Winterflood & Anne Butcher, June 10.
Dennys Elye & Sarah Edwards, June 24.
James Kindall & Hannah ——, Aug. 28.
Richard Newman & Mary Pereman, December 20.
1642 John Willoughby, gent., & Margaret his wife were marryed April 20. [1649.
1647 John Newman, widdower, & Sarah Hogg, Sept. 29.

Space of nearly half a page blank, and the first entry on the next page is dated 1654.

BURIALS.

[83]
1560 Simon Eesborow was buried the 6th. day of April in the yeere of O. L. God 1560.
John Purkas, 7 februarie.
1561 Elioner Maught, 13 februarie.
Christopher Harvie, 21 februarie.
Margaret May, 23 februarie.
Thomas Harvie, 28 februarie.
1562 Agnes ffiche, 12 April.
Agnes ffiche, the daughter, 23 May.
1563 Alice Berd, 30 December.
Anne Plomb, 5 Januarie.
1564 Agnes ffrench, 6 April.
Barbara Tongue, 22 June.
Thomas Plomb, 2 Septeber.
John Rizing, 17 Septeber.
Richard Yeldam, 25 Septeber.
Robert Pollard, 1 October.
John Thètford, 22 Deceber.
Rose Edward, 20 Januarie.
1565 Alyce Gaskyn, 13 October.
William Tongue, 4 Januarie.
Alyce Garroulde, 25 Januarie.
1566 Elizabeth Tyttrylle, 28 March.
Anne Farthing, 17 August.
Elizabeth Rawlinge, 23 Septeber.
John Hayward, 4 Noveber.
Katherine Gridley, 23 Januarie.
John Pollard, 9 Februarie.

Thomas Plomb, the elder, 19 Februarie.
1567 Henrie Bust, 1 April.
Joane Butcher, 2 Julie.
Richard Underwood, 11 August.
William Hamont, 3 October.
William Humfrie, 28 October.
Elizabeth Adcocke, 9 Noveber.
Joane Hulle, 18 Februarie.
1568 Anne Browne, 26 March.
John Mote, s. William Mote, last day of March.
Briget, d. Heugh Rawling, last day of March.
Alyce, d. Thomas Hybys, 14 April.
Katherine, d. Thomas Hybys, 17 April.
Barbara, d. John Greene, 23 April.
Richard, s. Henrie Thelford, last day of April.
Ane, the wife of Nicholas Warde, 8 November.
Elizabeth, d. William Edward, 2 Januarie.
Elizabeth, d. Henrie Reade, 16 Februarie.
1569 Elizabeth, d. Thomas Adcocke, 18 April.
John, s. John Harrington, 17 August.
Joane Adams, widowe, alias Butcher, 14 October.
Plesance, d. Edward Richardson, 21 December.
1570 Ane, d. Christopher Fitch, 3 Septeber.
Grace, d. Thomas Cracherood, 15 September.
John Craneford, servant, 6 October.
Edward, s. Christopher ffitch, 21 Januarie.
Richard Perrye, of flowers Hall, 23 Januarie.
Alyce, d. Henrie Snellocke, 1 ffebruarie.
Katherine, d. John Harrington, 19 Februarie.
Edmund, s. Robert Turner, 16 March.
1571 Richard, s. Richard Mote, 12 Maye.
John, s. William Mortemer, last day of Maye.
Ane, d. William Mortemer, 18 June.
Samuel, s. William Edward, 22 June.
Peter, s. William Stebbinge, 21 Julie.
Agnes Rizing, widowe, 29 December.
George, s. John Bust, 3 Januarie.
John, s. John Edward, the elder, 13 Januarie.
[84] Alyce, the servant to Robert Briant, 28 Januarie.
Henrie Reade, 16 Februarie.

1572 Joane Mortemer, wid., 26 March.
Joane, w. John Skeltie, 3 June.
Steven Tittrill, 3 Julie.
Thomas, s. John Cozin, 8 September.
Margerie, w. James Snellocke, last day of October.
Marie, d. Thomas Cracherood, the elder, 20 february.
1573 John Odye, servant, 3 May.
Phillys, d. Richard Everide, 18 May.
Nichobas Warde, wid^e^., 18 August.
William, s. Hugh Rawling, 20 August.
Richard, s. Paule Rawling, 9 September.
ffrancis, s. Richard Yeldam, 24 September.
Rose, d. Thomas Spiltimber, 4 November.
Margaret Gips, wid., 8 December.
1574 Joane Smitton, wid., 14 April.
Alyce, w. William Reade, 14 Maye.
Richard, s. William Edward, 25 October.
John Reade, the bas: 12 Deceb.
1575 John, s. Richard Hulle, 26 April.
Alyce, w. William Stebbing, 24 Julie.
Robert, s. William Reade, 24 September.
Thomas Grene, 12 Novemb.
1576 John Cracherood, 3 Januarie.
Thomas Caunts, 15 Januarie.
Robert Bryne, 15 Februarie.
Nicholas Belle, servant, 10 Maye.
Thomas Spiltimber, 12 June.
Alyce, w. William Pollard, last day of June.
1577 Thomas Pollard, 2 Julie.
Elizabeth Sanders, servant, 30 October.
Anne, d. Henrie Smith, 1 November.
William Bateman, 1 Deceber.
Christopher Fitch, 18 Januarie.
1579 John Hamond, 20 June.
1580 Henrie, s. Henrie Smith, 29 Maye.
Joane Pollard, widowe, 1 November.
Elizabeth Titrell, widowe, 4 November.
ffrances Humfrie, 6 Noveber.
Henrie Harrington, 17 Noveber.
Richard Denis, a stranger, 26 November.

1581 Margaret, d. Thomas Bateman, 25 Julie.
John Browne, the elder, 9 August.
Samuel Bigg, 24 August.
Margaret Pollard, 20 November.
Richard Yeldam, 27 Noveber.
Henrie Thetford, 27 Februarie.
1582 Ambrose Sparowe, 12 April.
[85] Barbara, w. Henrie Snellock, the elder, 27 June.
Elizabeth, d. Henrie Snellocke, the elder, 14 Julie.
James, s. John Harrington, yeoman, 7 September.
John Francis, 4 Noveber.
1583 Anne, w. John Bosall, 19 May.
Anne, w. Thomas Bosall, 11 June.
Anne Briant, widow, 20 June.
Christian, d. Robert Edward, 7 August.
Widowe Constable, 15 August.
1584 Elizabeth, w. John Harrington.
Elizabeth Richardson, 24 August.
Alyce, w. Richard Mote, 4 Septeber.
The wife of Edward Hampton, 22 November.
Thomas Edward, 17 Deceber.
John Edward, 29 Januarie.
The wife of William Pollard, 1 Februarie.
1585 Joane, w. William Mortimer, 3 April.
Alyce Spiltimber, 16 April.
Alyce Browne, widow, 1 May.
Margerie, w. William Harrington, 24 May.
Frances Barnes, 5 June.
John Bacon, 2 November.
William Mortimer, 3 November.
Henrie Bateman, 8 November.
William Cracherood, the elder, 12 Januarie.
William Bigge, 26 Januarie.
1586 Roger Brewster, servant, 30 April.
John Plume, 1 October.
Richard Snellocke, 1 Deceber.
Anne, w. Henrie Smith, 12 December.
Anne Bever, 14 December.
Alyce, d. John Percy, 11 March.
1587 William Bosall, 15 Julie.

Joane, w. Robert George, 8 September.
John Syke, last day of October.
Alyce Tittryll, 8 Noveber.
The wife of John Warman, 12 November.
John Hamond, 12 Noveber.
Thomas Whiting, 3 December.
Elizabeth Cracherood, widow, 15 februarie.
Mother Harrington, widow, 2 March.
1588 John Gipps, 27 Julie.
Katherine Tonge, widow, 12 Januarie.
Joseph, s. Lewes Bret, 15 Januarie.
[86] John Bust, 10 Februarie.
John Pollard, 13 March.
Robert Cartwright, 27 April.
Thomas, s. Henrie Smith, 8 May.
Marian Reade, widow, 10 August.
——— Tylbroke, 11 Noveber.
Matthew Edward, 27 Noveber.
Elizabeth Bosall, 1 March.
1590 Edee Easkin, w. Richard Easkin, 13 May.
Katherine ffrancis, widow, 2 June.
Sarah Hamond, d. John Hamond, 29 Januarie.
Anne Hamond, 12 March.
1591 John Browne, 2 Julie.
Thomas Bateman, 10 Julie.
William Bocher, 10 October.
Julian Adcoke, 20 October.
John Bird, of flowers hall, 28 November.
ffrancis Bird, 13 Deceber.
1592 Richard Easkin, 29 Septeber.
Sarah Turner, 7 Januarie.
A poore pedler man, 7 Januarie.
The two yonge daughters of William Edward, of Bradfields, were burid the fifth day after their birth.
1593 Anne, d. Widow Greene, 18 May.
Rose, w. Henrie Snellock, the younger, 14 Julie.
Katherine Thorowgood, 24 August.
Alyce, w. Michael Tongue, 28 September.
Joane, d. Thomas Cracherood, 20 October.
A daughter of Thomas Hurrell, 20 October.

John Bayle, miller, 1 Deceber.
Sarah Tyler, 17 December.
1594 Henrie, s. Richard Mote, 19 August.
Joane, w. William Bocher, 22 August.
Alyce, w. John Cozen, 15 September.
William, s. William Thorogood, 3 December.
Henrie, s. John Reade, 25 December.
William Allin, last day of January.
Joseph, s. Ishak Cornwell, 17 March.
1595 John Harrington, 2 April.
Robert Lambe, a stranger, 19 April.
The wife of Robert Tongue, 19 May.
Elizabeth, w. William Edward, 28 May.
John Posler, 3 June.
William Reade, 5 June.
[87] ffather Kempe, 28 June.
Jerome Turtell, 5 December.
Margerie Bosall, 5 December.
The child of Lewes Brett, 7 December.
Widow Bayle, 11 March.
1596 Margaret King, widow, 17 October.
A poore begger woman, 26 December.
Thomas Buttall, 3 Januarie.
Cicelye Bust, widow, 10 Januarie.
Marie, d. Henry Snellock, 25 februarie.
John, s. John Fiche, 6 January.
1597 Thomas Smillon, 1 May.
The widow of John Browne, 10 June.
John Cosen, 28 Julie.
Alyce, w. Richard Mote, 29 August.
John Humfrie, 6 September.
Angela Byford, 10 December.
Elizabeth Coe, 14 December.
Richard Cowlet, 2 Januarie.
William Turtell, 12 Januarie.
Alyce, w. George Rule, 15 Januarie.
Margaret, w. John Hewes, 16 Februarie.
William Whiting, parson of Toppesfield, 22 Februarie.
Susan, d. Edmund Whiting, 1 March.
Alyce Rysing, 19 March.

1598 Alyce Greene, widow, 25 March.
Anne, d. William Bocher, of Gainsforde, 1 May.
Alyce Bosall, 13 Maye.
ffrancis, s. Henrie Snellocke, 28 Maye.
Margaret, d. William Harrington, 7 June.
William Harrington, 24 June.
Rose Cant, widdow, 14 Julie.
Alyce, w. William Edward, of ffullers, 11 August.
Margaret Hart, widow, 3 October.
Zacharias Smith, alias Annis, 26 December.
Denis, w. Henrie Snellocke, 5 Januarie.
[88] Marian Edward, widow, 18 Januarie.
Joane Edward, widow, 16 March.
1599 A childe of Thomas Howe (unbaptized) the last day of April.
Joane Bayly, a poore girle that dwelt in this towyne, 11 May.
Joane Bayly, w. John Bayly, 6 Julie.
Samuel, s. Edward Ostler, 12 July.
Elizabeth, d. Robert ——,* 26 July.
Susanna, d. Thomas Harvey, 22 August.†
1600 Mary, d. Jonas Spiltimber, 14 June.
John ffisher, 18 June.
Elizabeth, w. Henry Smith, 23 Julye.
Henry, s. William Cracherod, 24 Julye.
Sara, d. William Edwards, younger, 23 August.
Rose, d. Richard Edwards, 10 Septembe.
Hercules, s. widow ffisher, 29 Septembe.
Ellenor Cornell, widow, 30 October.
Steven, s. Steven Cant, 12 ffebruarie.
1601 John Bragg, 21 March.
Mary, d. widdow Fisher, 15 November.
1602 Richard Pollard, 28 April.
Mother Kempe, an aged woman, 24 July.

---

*Name not clear; it may be Cozan.

†Down to this point the register has been copied by the same hand as the baptisms down to this date; the next eight entries are in the same hand as the baptisms of 1600; the original hand-writing occurs again in the entries for 1601-4, but evidently making original entries and not a copy as the ink varies much in color.

Ann, d. Mr. Jonson, of Habridg, 2 August.
Mathew Thorogood, 2 November.
Henry Cirke, 20 November.
Mary Buttoll, 19 December.
Widdow Harrington, an ancient woman, 3 ffebruary.
Mother Linwood, 8 ffebruary.
1603 William Buttolph, 12 Aprill.
Isaac Hart, 22 May.
Henry Read, a child, 27 May.
Mary Read, 2 June.
John Bunting, 16 July.
Clement Boreham, 2 October.
John, s. Richard Edward, 9 October.
1604 Mother Bush, 23 August.
Mother Stapleton, 2 August.
[89] John, s. Richd. Edwarde, 5 October.
Symon Grene, 1 January.
Alse, d. William Batten, 19 March.
1605 Jone, w. Thomas Browne, senior, 27 June.
Ann Pettit, an ancient woman, over 100 years ould, 7 October.
Samuel, s. Robert Devorax, 10 December.
Margere, w. John Pollard, 22 December.
Ould Mother Seaman, 6 ffebruary.
Margaret, w. Thomas Hurrell, 9 ffebruary.
John Read, the elder, 21 ffebruary.
1606 John Baley, 29 April.
Mary, w. Jonas Spiltimber, 13 May.
Ould Mother Hedg, 11 September.
Alse, d. William Batte, 21 October.
Margaret, d. Henry Petite, 14 January.
Dorothye, d. Richard Edwardes, January 18.
1607 Mary, wiefe of John Brande, September 15.
Mother Gipps, Januar 12.
Thomas Browne, senior, february 28.
Susan, wiefe of Mr. John Cracherode, february 11.
William Smitten, March 11.
1608 Mary, d. Thomas Cooke, April 11.
Mary, wiefe of Stephen Cannte, April 24.
John Hedge, an old man, June 19.

Mary, wiefe of Thomas Bayly, May —.
Michael Richardson, June 26.
1609 Roberte, s. John Pollard, June 30.
William Caunte, July 2.
Joan, w. John Clerke, February 9.
Mary, w. Lawrence More, 22 March.
Thomas, s. Thomas Chatterton, 23 March.
1610 Thomas, s. Henry Laver, 4 Aprill.
Richard, s. Hercules Evans, 12 Aprill.
Ann Butcher, an ould woman, 20 July.
Dorathy, d. Thomas Brown, 2 August.
Richard Everase, 21 August.
Joan Barber, an ould woman, 21 August.
William, s. James Shull, 23 September.
Robert Georg, 24 October.
Anne Kemp, 11 November.
Abraham Humpye, 4 December.
Ann, d. Clement Boreham, 19 January.
1611 John Read, of the hill, 22 June.
ffrancis, w. Thomas Plumb, 1 September.
1612 William Edward, sen., 22 March.
Thomas ffitch, 23 May.
Dorathye, w. William Butcher, 12 June.
Thomas, s. Thomas Browne, 23 Julie.
Margaret, d. Lowrance Moore, Jan. 18.
Henry Snellocke, Jan. 28.
John Pollerd, March 22.
1613 John Parker, Aprill 2.
Widdowe Cirke, April 20.
Rose, w. Richard Edwards, May 16.
Henry Smith, the elder, May 18.
William Massy, Aug. 18.
Thomas, s. John Perry, Nov. 21.
Susan, d. William Werhead, Jan. —.
Samuel, s. William Edward, Feb. —.
1614
[90] April 7. Anne, d. Richard Bucher.
April 18. Rebecca, d. Daniell Dod.
May 9. Robert Perry.
May 20. John Heart.

July 29. Joane Smitten.
Sept. 19. Oliver Stebben.
November 27. Elizabeth, w. Jerimie Pearman.
March 16. Ede, the base daughter of Adwy Fisher.
March 4. Phillis Pollard, widdowe.
1615 June 13. Hennery Cant.
July 24. Hennery, s. Hennery Pettit.
Sept. 3. Alice, w. William Read, the elder.
Sept. 12. Robert, s. John Browne.
Sept. 24. Rose Bacon, single woman.
Sept. 29. Adwy ffisher, single woman.
Dec. 7. Joice Spiltimber, single woman.
1616 May 9. Sarah, d. Josias Pollard.
May 11. Alice Harrington.
May 14. John Cousin.
July 21. Ellen, d. Jeames Shedd.
June 12. Anne, d. William Sparke.
Aug. 3. Joane ffytch, widow.
Aug. 18. Mathew, s. Hennery Bevis.
Sept. 9. ffrances, w. William Edwards, jun.
Nov. 27. Henry Smith.
December 26. Thomas Gardiner, the elder.
January 1. Thomas Edwards, of Bradfields.
January 25. Susanna, d. Richard Gippes.
March 2. Elizabeth, d. Giles Elsing.
March 2. Thomas, s. Richard Raven.
1617 March 28. Alse, d. Richard Raven.
May 10. Ellen Fisher.
Oct. 26. John, s. John Smythe.
Dec. 23. The wife of Hercules Evans.
Dec. 27. George Rule.
Jan. 8. Mary, w. Robert Edwards, the elder.
Jan. 23. William, s. Henry Bayley.
Jan. 28. Rebecca, d. Daniell Dod.
Feb. 1. Robert Warner, the elder.
Feb. 28. Jeremie Amie, of Abingen in Cambridge-[shire.
March 7. William, s. ——— Wallis.*
March 23. William Paine.

*Illegible; all the writing here is very bad, the letters not being carefully formed.

1618 June 14. Thomas Cracherode, gent., the elder.
Aug. 3. John, s. Thomas Mathew, alias Miche.
Sept. 4. Elizabeth, w. Thomas Harvy.
Nov. 3. Thomas, s. John [Gore ?].
Jan. 30. John Lampson, son-in-law to David Plaile.
February 2. Marie Buttall.
Feb. 25. William Read, senior.
[91] March 1. Roger Hayward.
March 19. Joane, w. Samuell Hamond.
March 23. Marie, w. Roger Edwards.
1619 March 31. Richard, s. Thomas Wight.
Aug. 20. Margaret, w. John Bust.
Nov. 28. Anne Bayly.
Feb. 9. Joane Rule, widdow.
1620 May 12. Ellen, w. John Cratchrode, gent.
July 9. Richard, s. Richard Read.
July 10. Alse, w. Edward Moore.
July 11. Daniel Dod.
July 17. Robert, s. William Read.
July 20. Alse, d. William Bateman.
Aug. 9. Henry Bevis.
Aug. 22. Mathew Bateman.
Sept. 1. William, s. William Spark.
Sept. 1. Tomazin, w. John Knoxs.
Sept. 3. Sara, d. William Boram.
Oct. 28. Robert Harrington.
Novemb. 24. John, s. John Ridgewell.
Dec. 4. Widdow Harrington, sen.
Dec. 29. Robert Pratt.
Feb. 6. Anne Gardiner.
Feb. 18. Henry Lidmore.
Feb. 25. Anne Evered, wid.
March 7. Edward Moore.
March 14. Saunder Buckley. [this place.
1621 April 20. Richard King, Dr. of Divinity & Rector of
John Cratcherood, gent., 6 July.
Amy, w. Jeremy Parmenter, 19 August.
Margaret, d. Adler Newman, 20 September.
Joanne ffinch, 19 November.
Ambrose Thompson, 10 December.

1622 Thomasin, w. Jeremy Pearmaine, 10 May.
[92] Mary, w. John Buckly, 26 May.
ffrancis, w. Samuel Edwards, 7 June.
Elizabeth, w. Richard Bateman, 2 July.
Joseph Marriner, 18 Sept.
John, s. John Drury, 7 November.
William, s. Joseph Marriner, 22 December.
[Samuel ?] s. Richard Edward, 9 March.*
1623 Tamosin, d. Jeremy Pearmaine, 19 June.
John Read, 23 June.
Mary Claidon, 23 July.
Susan, d. John Redgwell, 4 August.
Samuell, s. Samuell Symons, 4 November.
William, s. William ffitch, 28 November.
Jane, d. William Pamplin, 23 January.
Margaret Read, widdow, 10 March.
1624 Susan Mantt, 9 May.
Allice Batty, widdow, 3 June.
Allice, w. John Start.
Margarett Borum, widdow, 28 June.
William Simpson, 13 August.
Daniell, s. John Dod, 21 September.
Anne Hornsie, 5 October.
Robert Warner, 15 October.
Edward Ostler, 24 October.
Elizabeth Poole, 14 December.
[93] Samuell Smith, 5 March.
William Cracherode, gent., 10 March.
1625 William, s. Thomas Buttoll, 23 Aprill.
Ellenor Harvy, 15 May.
Richard Butcher, 24 June.
Agnes, w. Robt. Wankfourd, 4 July.
Agnes, d. Robt. Wankfourd, 19 August.
Allice, w. William Reade, 24 August.
Brigett, w. John ffisher, 27 August.
Mary, w. Thomas Greene, 8 September.
Thomas Chadderton, 15 September.
Joane, w. John Starte, 28 September.

*Partly erased, it is a baptism; there have been entries made and erased so as to be illegible down to the bottom of this page.

Mary, d. Raph Sewell, 5 October.
Thomas Baily, 24 October.
Elizabeth, w. William Bryant, 27 October.
Anne Butcher, 27 November.
Margaret South, 30 November.
Allice Evans, 30 December.
John Redgwell, 24 January.
Elizabeth, d. William Batty, 21 ffebruary.
Mary, w. Henry Paine, 25 ffebruary.
1626 William ffarrar, 25 March.
Thomas, s. William Levet, 25 March.
Roger Edwards, 26 March.
Sarah, w. Thomas Gardner, 4 Aprill.
Margarett Greene, widdow, 2 May.
Thomas, s. Robert Pollard, 21 May.
[94] John Bryant, the elder, 12 June.
William Johnson, 1 July.
Marie Allen, 5 August.
Anne, d. Thomas Trapnell.
Margarett, w. Henrie Clark, 5 August.
George, s. George Gyps, 18 August.
William, s. Roger Hoyden, 2 September.
Margarett Read, widdowe, 9 October.
Robert Edwards, the elder, 24 October.
Allice, w. Richard Paine, 30 October.
Susan, w. John Mising, 1 November.
Susan, d. Michael Richardson, 4 December.
Henrie, s. Henrie Paine, 29 December.
Thomasin ffuller, 29 January.
Grace, d. John Pollard, 7 March.
John, s. John Purchas, 18 March. [2 Aprill.
1627 A wandering beggarman whose name was unknown,
Susan Cooke, 9 Aprill.
Rose, w. John ffisher, 15 Aprill.
Thomas, s. John Dod, 19 May.
Susan, d. Thomas Paynell, 28 June.
Thomas, s. William Bryant, 6 July.
Robert Edwards, of Comans, 19 July.
William, s. John Start, 24 August.
Robert, s. John Perrie, 6 September.

Joseph Bragg, 2 October.
Thomas, the base son of Marie Butcher, 27 October.
William Smith, of Graies, 22 November.
Thomas Mathew, 11 December.
Margerie, d. John Start, 12 December.
Allice, w. Richard Raven, 20 December.
[95] Susan, d. Rafe Sewell, 31 December.
Henrie Paine, 10 Januarie.
Grace, d. Widdow Battie, 15 March.
Judith, d. Henrie Bailie, 16 March.
1628 Joane, d. ffrancis Kendall, 24 June.
Marie, d. Robert Edwardes, the elder, 25 June.
John Start, the elder, 26 June.
Agnes, d. ffrancis Kandall, 19 Julie.
Robert, s. Ralfe Sewell, 19 Julie.
Henry Browne, 3 August.
Rafe, s. Rafe Sewell, 23 August.
Susan, w. William Levett, 16 September.
Rachell, w. John Gyps, the elder, 7 November.
Anne, w. John Mising, 7 November.
Elizabeth, d. Thomas Trapnell, 8 November.
Marie, w. Rafe Sewell, 8 Januarie.
William, s. John Perrie, 14 Januarie.
Daniell, s. John Busie, 12 ffebruary.
1629 Henrie Bailie, 10 Aprill.
Richard Paine, 28 May.
Marie, w. William Chadderton, 24 June.
John, s. John Drury, 23 August.
Thomas, s. Thomas Harvie, 2 October.
Marie, d. Roger Hoyden, 23 October.
Robert Hog, 29 December.
John Bryant, 29 January.
Anne, w. John Bellowes, 10 ffebruarie.
Sarah, w. Josias Pollard, 6 March.
1630
[96] ffrancis, d. Edward Clay, 16 Aprill.
Samuel, s. William Edward, of ffullers, 1 May.
Robert Siblie, 13 May.
Jone Bull, 26 May.
Marie, d. John Laver, 7 June.

Anne, d. Edward Clay, 26 June.
Elizabeth, d. Thomas Harvie, 3 July.
Anne, d. John Busie, 5 July.
Margarett, d. Edward Clay, 29 July.
Anne Harvy, 30 July.
Robert, s. Samuell Symons, 2 September.
John Buckly, 1 October.
Jone Ellis, 12 October.
Allice, w. —— Edward, of ffullers, 9 November.
Sarah, d. Steven Warner, 10 November.
David Warner, the elder, 11 Januarie.
Daniell, s. Barnard Sibly, 23 January.
Luce Houchin, widdow, 6 ffebruary.
Adler Newman, 15 ffebruary.
William, s. Thomas Emsden, 16 ffebruary.
Thomas Browne, 27 ffebruary.
Marie, d. Robert Ellis, 23 May.
1631 Margerie Titterell, 27 May.
John Bust, 7 August.
Arthur Winterfloud, 9 August.
Richard, s. Peter Hale, 22 August.
John Amys, 11 September.
[97] Rose, w. John Perrie, 14 September.
Agnes Smith, 5 October.
John, s. John ffitches, the younger, 16 October.
Alice Newman, 22 October.
Susan, d. John Bellowes, 12 November.
ffrancis, d. John Busie, 14 November.
Richard Paine, 20 December.
William Cosin, 25 January.
Philip Poole, 10 ffebruary.
Marie, w. Samuell Bateman, 17 ffebruary.
ffrancis, d. Samuell Edwards, of ffullers, 29 ffebruary.
Elizabeth Cosin, 7 March.
1632 William, s. Samuell Edwards, 30 April.
Elizabeth Spurge, 25 May.
William Greene, the elder, 27 June.
Samuell Bell, 8 July.
Anne, w. William Butcher, 3 August.
Marie, d. John Bryant, 10 August.

Thomasin Bailie, 16 August.
Rachell, w. William Edwards, of ffullers, 15 October.
Margarett Farrar, widdow, 20 October.
[98] Susan, w. John Gyps, the elder, 25 ffebruary.
1633 Hester, d. Robert Warner, 11 June.
Steven Cant, 10 July.
1634 Susan King, 8 June.
Roger Symons, 10 June.
Anna, d. Anna Scooline, 26 August.
Elizabeth Titterell, 19 September.
George Trapnell, 28 September.
Margaret Beard, 7 October.
Peter Argent, 16 November.
Henrie Bateman, 21 November.
Anne Symons, 27 November.
Anne Titterell, 28 November.
Marie Redgwell, 8 ffebruary.
Henrie Pettitt, the elder, 12 ffebruary.
Dorothy, w. Edmund Cocksedge, 12 ffebruary.
1635 William Pamplin, 18 Aprill.
Richard Titterell, 26 July.
Thomas Read, 16 August.
Lydia Edwards, 4 September.
[99] Lydia Wade, 23 September.
William Edwards, 25 September.
Thomas Browne, 3 October.
Peter Coote, 12 November.
Marie Paul, 4 December.
Anne Smith, widdow, 11 ffebruary.
Marie, w. Nathanaell Paul, 25 ffebruary.
1636 Elizabeth Perrie, widdow, 29 August.
William Bryant, 16 June.
Susan, w. Mr. Thomas Cracherode, 22 June.
Dorothy, w. Mr. Samuel Simons, 3 August.
John Southy, 18 August. [ber.
William, s. Robert Edwards, the younger, 18 Septem-
Grace Ostler, widdow, 20 September.
Sara Pamplyn, widow, 25 September.
Anne Cirke, a child, 19 October.
Anne, d. William Berd, a child, 5 November.

Dorcas, d. John Laver & Mary, his wife, the last day of ffebruary.
Susan, d. Thomas Harvey & Susan, 7 March.
Mary ffisher, widow, 9 March.
John Gyps, widower, 12 March.
Samuel, s. Thomas Emsden & Elizabeth, 12 March.
1637 Robert Maltiward, s. Robert Maltiward, gent., 14 April.
[100] Susan Pettit, widow, 29 May.
Elizabeth, d. Michael Richardson, 24 June.
Elizabeth Monke, widow, 29 June.
Barbara Laver, w. Henry Laver, 26 August.
Joane ffinch, w. John ffinche, 10 September.
Thomas Warner, 30 October.
Clemence Clerke, w. Henry Clerke, 3 November.
John, s. William Murken & ffrancis, 4 November.
Judah Brewster, w. Robert Brewster, 13 December. [Robt. written and erased, Judah written over.]
Hercules Eveins, 25 December.
Alice ffitches, w. John ffitches, 27 December.
Mary Mumford, w. John Mumford, 14 January.
Jeremy Pareman, 14 February.
ffrancis Simson, s. John Simson, 24 Jan.
Rose Cooke, d. William Cooke, 29 January.
Henery Clerke, widower, 6 Feb.
The widow Bryant, 12 March.
1638 William, s. Edward Clay & Anne, 7 April.
Alice Cosin, widow, 7 June.
Katherine Chaterton, w. William Chaterton, 26 July.
Elizabeth Earle, w. George Earle, 26 July.
Allice* [?] w. John Perry.
Thomas Harvy, 19 Aug.
Mary South, 24 Aug.
Mary, w. John Reade, 15 Sept.
Margaret, d. Robert Pillow, 17 Sept.
Susan, w. John Gipps, 5 October.
Old More, 18 November.
William, s. Edward Taylor & Elizabeth, 20 December.
Dorothy, w. William Browne, 26 June.
William, s. William Redgewell and Joane, 9 feb.

*Name written over an erasure and almost illegible.

Anne Edwards, w. William Edwards, 21 february.
1639 John Chambers, a child, 14 April.
John, s. John Bryant, 6 June.
Mary, d. William Browne, 10 June.
Anne Bowyer, 29 June.
George Earles, s. George Earles, 4 July.
Sarah, w. Henery Pettit, 15 July.
John Bowtell, 20 July.
Samuell, s. Samuell Bateman, 22 July.
[101] Thomas, s. Thomas Emsden, 25 July.
Thomas Mumford, 28 July.
Sarah, d. Henry Pettit, 18 August.
Elizabeth Edwards, widow, 19 August.
Robert Edwards, s. Richard Edwards, 21 August.
Alice, w. Robert Edwards, senior, 1 September.
John, s. William Reade, October 20.
Susan Newman, widow, October 24.
William, s. William Berd & Sarah, Oct. 30.
Elizabeth More, widdow, Nov. 12.
Christian Overed, w. John Overed, clerke, Nov. 20.
——— Amye, widdow.
Jonathan, s. Thomas Roote & Anne.
Sarah, w. William How, Jan. 12.
Martha, d. Michael Richardson, 8 March.
John Quie, the same day.
Thomas Edwards, March 10.
Robert, s. John Laver & Mary, was buryed March 15.
ffrancis Gall.
1640 Joane, w. William Redgewell, April 4.
David Warner, April 6.
William, s. William Redgewell & Joane, April 20.
Judah Pollerd, June 17.
Anne Busye, w. John Busye, June 21.
Anne, baseborn child of Anne Winterflood, June 28.
Anne, w. Thomas Warner, July 25.
Ellen, w. Robert Pollerd, Sept. 12.
Robert Harrington, gent., Sept. 27.
Elizabeth, d. Edward Tayler, Novemb. 28.
Robert Pollard, January 14.
———, s. Robert Pollard, feb. 1.

——— Warner, widow, feb. 22.
1641 Michael Richardson, the last day of May.
Susan, w. Sydney Eivens, June 11.
Margerie, w. William Boreham, June 21.
———, d. Robert Warner, Sept. —.
ffrancis, d. John Briant, Nov. 24.
Henery Laver, the elder, December 7.
William Butcher, Dec. 14th.

This page is filled for nearly all of its length; it is followed by a page completely blank; the next page is dated 1655.

NOTE by the copyist.—Names found in this register which yet are borne by present (1902) inhabitants of Toppesfield and neighbourhood.

Allen (Alwin), Argent, Barber, Brewster, Bromley, Butcher, Clarke, Coote, Earl, Ellis, Fitch, Gurteen (Gurton), Hale, Hall, Houchin, Mumford, Newman, Pannell (Paynell), Parmenter, Purkis, Read, Ridgewell, Ruggles, Seaman, Shed, Smith, Sparrow, Wade, Yeldham. There were also Eleys here till quite recently.

There are houses called: Thurstons, Flowers Hall, Quays, Peacocks, Mortimers, Hurrells, probably after people named in these lists.

ST. MARGARET'S CHURCH, TOPPESFIELD, ENGLAND.

A SKETCH OF

TOPPESFIELD PARISH, ESSEX CO., ENGLAND,

BY REV. H. B. BARNES,

Rector of St. Margaret's,

— AND —

THE HISTORY AND ANTIQUITIES OF

TOPPESFIELD PARISH, ESSEX CO., ENGLAND,

BY PHILIP MORANT, CHELMSFORD, 1816.

Annotated and Edited

BY GEORGE FRANCIS DOW,

Secretary of the Essex Institute, Salem, Mass.; Secretary of the Topsfield (Mass.) Historical Society; Member of the American Historical Association.

Reprinted from the Topsfield, Mass., 250th Anniversary Proceedings.

The Merrill Press,

TOPSFIELD,

1900.

A SKETCH OF TOPPESFIELD PARISH,

ESSEX CO., ENGLAND.

BY REV. H. B. BARNES, RECTOR OF ST. MARGARETS.

For the last six months I have been trying to gather material for a sketch of the history of ancient Toppesfield. The work would be by no means easy even for an expert, for there appear to have been no previous workers in this field, from whom to gather without toil that which must in the first instance have been discovered at the cost of much time and labour.

Of course the chronicler has the old records on the tombs, the old account books, as well as the old registers, which he can always consult, and which probably would reveal tales of deepest interest to any one who has leisure to study them, and experience and skill to understand the meaning of that which is written in these old-world records, but the present writer confesses with sorrow that even had he the time to spare he has not got the skill; but he hopes that he is no dog in the manger; so should any one (and especially any one interested in the connection between Topsfield and Toppesfield) wish to work up all that can be learned from these original documents, he may count on being met with the heartiest welcome, and the fullest help that can be rendered.

As then, (in the absence of other men's writings from which to steal, and of ability to make original researches) it is impossible to write any account of ancient Toppesfield which shall not be of an imaginative rather than an historical character. I have thought that perhaps some short account of the Toppesfield of to-day might be of interest.

The village is situated in the north-eastern corner of the County of Essex, near to the borders of Suffolk on the east, and of Cambridgeshire on the north; the country is not by any means of the level character that is usually attributed to the whole of Essex. There are no great hills but there is no flat country; all is undulating. Toppesfield itself—whatever the origin of its name—certainly by its position deserves its designation; the church does not stand on the highest ground in the parish, but yet its tower serves for a land-mark for miles around, on all sides except the west, on which side a wood screens it from view; while in the parish about two miles in a southerly direction from the church, is found the highest point in this part of the county, excelled in the whole county only, if at all, by Danbury Hill near Chelmsford.

The soil is almost uniformly clay, and very good for wheat growing, and its fertility is such that even in the present time of agricultural depression there is not an unoccupied acre in the parish. Yet it must not for a moment be supposed that Toppesfield has escaped unscathed; very far from it. Thirty years ago it was as rich and prosperous a little place as could be found; now it is miserably poverty-stricken; then, there were numbers of well-to-do farmers, now, the land is farmed in large holdings by men who, for the most part, live in neighbouring villages; then, many of the old houses dotted about the parish were occupied by large and thriving families; now, the families have gone and many of the houses are either occupied by labourers (*e. g.* Olivers, Cust Hall and Fry's Hall) or are falling into decay as "Mullows" has done. The impossibility of making a living off the land, has driven the descendants of sturdy yeomen to seek elsewhere, the livelihood which the ground their fathers tilled, can no longer afford them.

Nor is the lot of the labourer better than that of the

farmer; though the cause of the trouble is in his case different; for farm labourers wages, have this year stood higher than they have ever been known to be before. But in the old days the daughters and wife would earn more than the father, and would do so without being necessarily taken away from home; even thirty years ago, straw plaiting was a great industry in this part of England. Old crones maintained themselves in comparative comfort by holding "schools" in which infants of quite tender years were taught to plait, and, as the children grew up, they plaited as they stood in their cottage doors or as they lolled about the roads, and their work was every week collected by higglers who came round for the purpose. All this has come to an end now; no straw plait is made here for it can be more cheaply imported from the East than it can be made at home; and though the money that was earned in this way is much missed, yet the village is happier and better for the loss of this business, for straw plaiting always seemed—wherever it was done—to bring a moral deterioration in its train.

There is however an indirect way in which the agricultural depression seriously affects the labourer; it makes it very difficult for him to get a decent cottage. The profits of farming having been so much reduced, the farmers have been unable to pay anything like the old amount of rent and this has hit the land-owning class very hard; in some cases the depreciation of the value of land has been so great that its capital value now is little more than its old annual rent; plenty of good land can now be bought for £7. an acre and in this price are sometimes included farm houses and outbuildings and cottages which have quite recently cost more than now they can fetch, even with the freehold of the land thrown in; small pieces of land without buildings fetch (except for some special reason) even lower prices. I heard last week of thirteen acres of good land in an adjoining parish being sold for no more than £40.

The landlords then, being so hard hit in all cases, and sometimes having positively *no* balance left after they have paid the "charges" on the estate (doweries it may be or pensions determined upon during the fat years of prosperity) are unwilling, even when, through having other sources of income, they are able, to spend more money than can be helped, on the up-keep of their farm buildings and the cottages on their farms; hence on every side the barns and out-buildings are more or less dilapidated, (though it must be owned that in this respect there has been a considerable improvement during the last two years) hence too the refusal to repair old cottages, so that cottage after cottage is condemned by the medical officer of health as unfit or unsafe for human habitation, and the inhabitants of the condemned cottages are obliged to seek their living elsewhere than in the old parish. As for new cottages, none have been built lately and none are likely to be built, for if the landlords cannot build them no one else will except from philanthropic motives, for it would be difficult to get a nett return of two per cent. on the minimum cost of erection.

The necessary results of such a condition of things are easily understood; the best of the young men go off to the towns, and there gain their living; many of them become policeman or employés on the railways; others become soldiers; the young women go out to domestic service and so the village is left with the old people and the young children to inhabit it. The proportion of the old is something remarkable; that the climate is extremely healthy and that longevity is much more common here than in most places, may have a little to do with it, but fails altogether to account for the wonderful proportion of old people in the population; no, the reason is that the young men and women as soon as they grow up go off elsewhere to seek a better market for their labour; and while we regret losing them, and fear that many of the men like the married man of the story find the change "none for the better and all for the worse," there can be no doubt that the course they take is the one which must seem most reasonable to those who have no knowledge of the condition of unskilled labour in the great towns. The extent to which this exodus is reducing the population of the parish may be judged from the fact that while in 1831 there were 1088 inhabitants; in 1881 there were 861; in 1891 790, and in 1901 there is no doubt that there will be a still further reduction. It is impossible to form an accurate estimate, but I should guess the number at 650, basing my calculation on the number of children on the school books, which is now 115, while in 1891 it was 146. I am glad to say, however, that the average number in attendance for this year is higher than it was then, for while in 1891 the average was 111, it is for the time that has passed since the beginning of the current school year on April 1st last* 113, which we are proud to consider would be a remarkable performance for any school, but which is highly creditable in a parish where some of the scholars live two and one-half miles away from the school door. The school is a voluntary school supported by a voluntary rate of 4d in the £1, in addition of course to the Government grant; the total cost for a scholar in average attendance being about £2. 10. 0. per annum; the buildings are good and roomy, and would accommodate nearly double the present number of scholars. In the school is also held an evening continuation school for young men which was begun this year and which has been doing fairly well. In this same building are held the meetings of the members of what is known as "the school club," an excellent Benefit Society, a branch of the National Deposit Friendly Society. The Toppesfield branch started some fifteen years ago by the then Rector, the Rev. C. F. Taylor, has over 100 members; many of them however are now living in distant parts and some come from neighbouring villages. Toppesfield has reason to feel proud of its school and of its Benefit Society.

Near the School is the church which is dedicated to St. Margaret; the tower looks imposing from a distance but when examined more closely proves to be a rather poor specimen of the architecture of the beginning of the eighteenth century; there was an old tower, the inside of which must have opened on to the church, with a lofty early English arch, and which is said to have been built of flint and rubble; this fell down on July 4th 1689, and was replaced by the present structure of brick; the tower contains five bells, two of which however need recasting. The church consists of a chancel, nave, and south aisle with a gallery at the west end, against the tower. The chancel contains an interesting old

*It is only fair to state, that during the months April, May and June, there were ten more children on the books, but the average weekly percentage of children present is, for this year, over ninety-five.

tomb surmounted with a cross, built half in and half out of the south wall. There is no inscription on the tomb, and it is not known to whom it belongs. In the floor is an old brass, bearing the figures of a man and woman, and with the inscription

> Pray for the sowlys of John Cracherowd and Agnes his wyff: the whyche John decesyd the yere of Our Lord God 1513, upon whose sowl Christ have mercy.

Near to this there is another brass plate with the inscription:

> Here lyeth buryed William Cracherod, Gent, who died Xth of January 1585, and Eliz: his wyfe the XVIIth of Feb. 1587.

Near to this again there is a tomb, with a full-sized effigy of a man, bearing no inscription, but probably containing an earlier member of the same family of Cracherod.

On the walls of the chancel are commonplace memorials of three former Rectors,* and two memorials of ladies which may be worth transcribing; on the north wall there is a marble monument bearing various symbolical devices† and this inscription:

---

*Against the east wall of the chancel is a small mural monument, upon which is written as follows:—Ego Richardus King, patria Herefordiensis, educatione Oxoniensi, pofessione theologus, officio capelloneus Jacobi Regis ferenissimi & hujus ecclesiae vicarius indignus, hoc in loco sacrosancto sponte depono & recondo corporis exuvias laus Deo, salus ecclesiae, & animae meae requies in aeternum. Amen. [For illustration of this tablet, see, The Ancient Sepulchral Monuments of Essex. By Frederic Chancellor, p. 325, London, 1890.]

In English:—I Richard King, by country an Herefordshireman, by education an Oxonian, by profession a divine, by office a chaplain to king James and the unworthy vicar of this church, willingly deposit my remains in this sacred place.—Praise be to God, health to the church, and rest to my soul for ever. Amen.—*History of Essex* (*Co.*). *By a Gentleman. Chelmsford*, 1771.

†Two Bibles serve the office of trusses, upon which are two rows of books, that instead of two pilasters support a neat pediment, in the middle of which pediment is a beehive, and under the hive is written *indultria dulcis*, meaning *sweet industry*. Over the hive is placed a dove, with the words *fida simplex* (imparting *simple fidelity*) written below it. Six of the books which compose the pilasters are labelled thus:—Sacrae medit; Soliloquia; Publ. Prec; Praxis Pict; Flores Prac; Psalmi.—*History of Essex* (*Co.*). *By a Gentleman. Chelmsford*, 1771.

VIEWS SHOWING THE WEST PORCH AND INTERIOR OF ST. MARGARET'S CHURCH, TOPPESFIELD, ENGLAND.

Sacrum memoriœ pientiss[m] fœminœ Dorcadi (sic) uxori
Guil Smyth armigeri; qui eam prius viduam Guil. Bigg triumq
liberor matre, ob modestia, pietate prudentia singulare
duxit; et in familia prosapia celebre traduxit; ubi multos
annos ille, spendidœ hospitalitatis et candoris, illa
solertiæ fideique matronalis exemplar; clara omnibusq
nobilib[s] œque ac infimis chara sui memoria reliqueru
Laudatiss[m] aviœ suæ, sacra senecta lectione, meditatione
bonisq operibus indefesse consolanti tandeq inter incredibilia
sanctissimæ animæ gaudia ultro in cœlu avolanti H. Bigg
nepos hisce symbolis parentat et lachrymis. Hoc pago educata.
nupta; Cressingœ, mortua, sepulta.

Obiit 1663. Dec. 18 anno ætat 76.*

---

*In English:—Sacred to the memory of that very pious woman Dorcas the wife of William Smith, esquire; who married her, when the widow of William Bigg and the mother of three children, for her singular modesty, piety, and prudence; and placed her in a family of great eminence; wherein, he was many years a bright pattern of hospitality and goodness; she, of diligence and conjugal fidelity; persons of every rank held her in great esteem: the memory of them was dear to all who knew them. H. Bigg makes an offering of this and of his tears to his much esteemed grandmother, who incessantly comforted her old age, by reading the holy scriptures, by meditation, and by acts of goodness; and who, at length amidst the inconceivable joys of a most pious soul, willingly winged her way to heaven. She was brought up and married in this town: she died and was buried at Cressing. She departed this life December 18, 1633, in the 76th year of her age. Beneath this inscription is the figure of a lamb placed upon a bible, upon which is written these words: *Biblia fides sacra*, which mean, *Faith in the Holy Bible*: on one side the bible is the representation of a bleeding heart, as figurativeof her feelings for the distressed poor: on the other side is that of an expanded hand; doubtless as a symbol of her readiness always to assist them. The whole is prettily designed, and executed in a masterly manner.—*History of Essex* (*Co.*). *By a Gentleman. Chelmsford*, 1771.

On the South wall is a memorial of a young lady of eighteen:

> Her disposition was mild and benevolent
> her manners gentle and simple
> and most respectfully obliging
> her sentiments enlarged and liberal
> her understanding clear and comprehensive
> enriched with an uncommon extent and variety
> of attainments, of which she was so far
> from making an ostentatious display
> that she seemed unconscious she possessed them
> nay, the degrading conceptions she unhappily formed
> of her own worth moral and intelectual (sic)
> were probably the source of insupportable sufferings
> "The brain too nicely wrought
> Preys on itself and is destroyed by thought."

One cannot but wonder whether the young lady overburdened by the marvellous talents of which she was unaware sought relief in suicide.

The South aisle has a fine old oak carved roof, the date of which can be determined (by the combination of the pomegranite and the rose found on it) to be about the year 1500. At the east end of the aisle there used to be a window with fine old glass, but it having been found necessary, some half century ago, to build a vestry out beyond the aisle, the glass in the window was removed and left about to perish! this is not the only loss—caused by neglect or ignorance—that we have occasion to deplore. At the east end of this aisle there can be seen on one side a piscina, showing that an alter once stood there, and in the other, high up in the wall, the entrance to the rood loft of which no other trace now remains. The font, which stands in the aisle, has no other interest than such as is derived from its great age. The body of the church has nothing to recommend it, the seats are mean looking and uncomfortable for use, the pulpit is commonplace, the west gallery (in which, in the good old days of even fifty years ago or less, sat the performers on the fiddles and the flutes) is Jacobean, but while all built of oak is faced on its pillars with carved oak; the great oak beams which span the nave are similarly cased, and unhappily neither they nor the roof are in a sound condition. The right of appointing the Rector rests with the Crown; there were here at one time both a Rectory (which then was a sinecure) and a Vicarage; but the Bishop of London, about 1454, finding that the Vicarage had become too poor to maintain a clergyman, united the Vicarage to the Rectory. There is still a piece of the Glebe land known as "the vicarage," which forms a memorial of the old state of things.

The names are known of all the clergy of the Parish since 1300:

DATE.	SINECURE RECTORS.	DATE.	VICARS.
	John Hardy.*		William (died)
1327.	William de Grytton.	1331.	Stephen le Parker.
	John Cory.		John Hokyngton.*
	William Noble.	1385.	William Lambeleye or Welton.*
	William Barret.		
1385.	Thomas Haxeye.*	1394.	John Cukkowe.
1386.	Thomas Banaster.*		William Mersey. (died)
1386.	William Gray.	1431.	Richard Pumpy.*
	Nicholas Manvell. (died)	1432.	John Scarlette.*
1446.	William Breden.*	1433.	William Meyr.
1452.	John Hambalt.		John Peteville.
1454.	William Parker.	1448.	Henry Huyton.

RECTORS.

	William Parker.	
1492.	John Edenham or Ednam, D. D.	Preferred. Dean of Stoke; Canon of St. Paul's; Master of Corpus Coll.
1504.	Thomas Fermyn. (died)	
1520.	Adam Becansawe.	Agent of Thomas Cromwell.
1551.	Thomas Donnell, B. D.	Deprived.
1553.	Cuthbert Hagerston, M. A.	
1554.	Thomas Havard.	
1556.	Richard Wynne.	
1559.	Thomas Donnell, B. D.	Restored. Prebendary of Lichfield.
1571.	William Redman, D. D.	Preferred. Canon of Canterbury; Bishop of Norwich.
1578.	William Whiting.	
1598.	Edward Graunt, D. D.	Canon of Ely; Sub-Dean of Westminster.
1601.	William Smyth.*	
1603.	Theodore Beacon, M. D.	
1604.	Randolph Davenport, B. D.	
1605.	Richard Kinge, D. D.	Chaplain to James I.

* Resigned.

RECTORS.

1621.	Richard Senhouse, D. D.	Dean of Gloucester; Bishop of Carlisle.
1624.	Lawrence Burnell, D. D.	Chaplain to Charles I.
1647-1661.	No rector.	Thomas Overhead intruded.
1661.	Clement Thurston, M. A.	
1662.	Nathaniel Ward, M. A.	
1662.	Edgar Wolley, D. D.	Bishop of Clonfert.
1664.	Richard Collebrand, D. D.	Dean of Bocking.
1674.	Robert Wild, M. A.	Chaplain of the Rolls.
1691.	Thomas Willett, M. A.	
1735.	John Hume, D. D.	Bishop of Bristol, Salisbury and Oxford.
1749.	Samuel Squire, D. D., F. R. S., F. S. A.	Dean of Bristol; Bishop of St. David's.
1750.	Henry Herring, M. A.	
1772.	George Pawson, L. L. B.	
1797.	Lord Henry Fitzroy, M. A.	Canon of Westminster.
1828.	George Henry Gooch, M. A.	
1876.	John Sherron Brewer, M. A.	

* Resigned.

Since the death of which distinguished man in 1879 there have been five other Rectors.

In the Church and Churchyard many of these worthies lie buried, but none of their memorial stones are worth copying. There is one stone however near the Tower which records that:

> Here lieth the body of
> Sarah Norfolk wife of
> Samuel Norfolk the younger
> who was cruelly murdered by
> her husband Septr. 24 1775 at
> a farm call'd Elms in this Parish
> in the 25th year of her age
> The said Samuel Norfolk
> confessed the fact
> was hang'd and desected

The Parish registers date back to 1558 and are in a good state of preservation and fairly legible to those who have mastered the difficulties of the old form of writing; there are also old account books dating back to 1662, and deeds of an earlier date.

On the first page of the earliest register is written in Latin and in English, the doggrel rhymes:

Advent wills thee to contein
But Hilarie sets thee free again
Septuagesima said thee nay
But eight from Easter says you may
Rogation bids thee yet to tarrie
But Trinity gives thee leave to marrie.

The baptisms, marriages and burials are entered in separate parts of the book but mistakes occur every now and then, so that a marriage is entered among the funerals.

Near the church stand the two village inns, the Chestnuts, and The Green Man, both of them picturesque in appearance. The Green Man is as quaint and old-fashioned as it is comfortable and well-managed. The host, Mr. Charles Seaman, has held his house for over forty years, and it is commonly said that there is not an hotel in any of the neighbouring towns for miles round where guests are made so comfortable or where a dinner so well cooked and served can be had.

Standing back in a park-like meadow is the old Manor House known as Berwick Hall; a nice comfortable house, with some old oak in it, inhabited by Mr. Charles Darby, whose family name has been known in Toppesfield for some three centuries at least.

Beyond the "Park" of Berwick Hall is the Rectory, part of which also is very old, dating back to the 14th century. There are traces of a moat round both Berwick Hall and the Rectory. Two years ago (1898) a very fine oak ceiling with large moulded beams, and an old oak doorway, were discovered in one of the rooms, having previously been covered up with plaister and canvas. The Rectory is very sheltered on all sides being enclosed by well-grown trees and with a large old Tithe Barn lying on its north side.

About half a mile from the Rectory on the road to Yeldham, stands "Olivers," with a beautiful approach through an avenue; it is now inhabited by two labourers; there is a panelled room still in an excellent state of preservation though the woodwork has been unfortunately covered with paint.

Toppesfield Hall, which like Olivers, belongs to Mr. J. M. Balls, stands on the other side of the Yeldham road; it is a comfortable modern house inhabited by Mr. J. F. Benson, one of the church-wardens, who is a nephew of the proprietor.

Bradfields is a picturesque house lying rather low, and in a rather dilapidated condition.

Gainsfords is another old Manor house about two miles from the church, occupied by Mr. C. Dean Darby, a son of Mr. Darby of Berwick Hall; it also has some nice oak.

Flowers Hall, about another mile beyond Gainsfords, is another nice-looking house, not very large, but with a wonderful range of out-buildings; it is now occupied by Mr. Clarke who with his family of active sons gets excellent results from some of the least fertile land in the parish.

I have given as fair a description as I can of the Toppesfield of today. What is its future to be? there is I think but little doubt. London is but fifty miles off, though thanks to the bad railway accommodation it takes two hours to get there. The Londoner is more and more developing a love for a country residence, and when the favourite counties of Kent, Surrey and Sussex get filled up, as they are doing already, those who like quiet will go further afield. Automobilism, or electric railways, will make travelling easy, and then this corner of Essex with its healthy climate, its quiet beauty, its fertile soil, its fine oaks and other trees will attract the class of persons who want a nice house and a few acres of land. Then land will again fetch in this district ten times what it fetches now; then there will be plenty of employment in stables, gardens and pleasure farms for the men who now flock into the towns. But this will not be in my day. But even now Toppesfield is a pleasant happy place with inhabitants who are not very fond of strangers, but who are essentially good-hearted.

TOPPESFIELD, ENGLAND.

FROM

HISTORY AND ANTIQUITIES OF THE COUNTY OF ESSEX

(ENGLAND), BY PHILIP MORANT.

CHELMSFORD, 1816.

This parish* was so called from some Saxon owner, named Topa, or Toppa. It is otherwise written in records—Toppesfend, Toppesford, Thopefield. In Edward the Confessor's reign, some of the lands here belonged to freemen, named Alestan; to Duua; to Got, &c., but, at the time of the general survey, part was holden by Eustace, Earl of Bologne, and his under-tenant, Bernard; part by one Ralph; and a considerable share, called afterwards Camoys-hall, by Hamo Dapifer.

These lands were divided, soon after, into the following maners:—The maner of Berwick and Scoteneys; Gaynesfords; The maner of Husees; Cust-hall; The maner of Camoys, and the maners, or reputed maners, of Flowers-hall, Gobions, Hawkeshall, and Bradfield. Most of these, if not all, are Duchy lands, and belonged to the honor of Clare.

---

*Is of large extent, fruitful in its soil, and pleasant in its situation, but not being a great thoroughfare, the roads hereabouts are in general heavy and narrow. The village is but small and rather mean in appearance. *History of Essex Co. By a Gentleman. Chelmsford*, 1771.

This parish extends northward to Great Yeldham; to Finchingfield on the west; southward to Wethersfield, and on the east, to the Hedinghams. Distant from Clare, five, and from London, fifty miles. The village is small, and none of the roads passing through this district being leading thoroughfares, they are in general narrow, and not in very good repair. The soil is a deep tenacious marl, retentive of moisture, and universally requires draining. *Wrights' History of Essex County. London*, 1836.

TOPPESFIELD. A. 3332; P. 861; Rectory, value £900; 2 m. SW. from Yeldham; B. 6. A pleasant, retired village on a commanding emi-

THE MANER OF BERWICKS AND SCOTENEYS.

They were separate at first, but have been long united, and took their names from their respective ancient owners, as will appear in the sequel. Berwick-hall stands a little way south-west from the church. The mansion-house and lands of Scoteneys lie near Yeldham, about half a mile from Berwick-hall. These two constitute the chief maner in this parish, though not the largest. In King John's reign, *Albrey de Wic*, or Wykes, held this estate, of the honor of Bologne, by the service of three parts of a Knight's-fee. He sold it to *Gerebert de St. Clere;* it being then called 84 acres of arable, 3 acres of meadow and pasture, 4 acres of wood, 45 pence rent of assize yearly, 49 days work, and ten hens. Part of the estate, viz.: 8 acres of arable, 5 of meadow, 4 of wood, &c., were holden of Ralph de Camoys.

Scoteneys was then distinct from it, and belonged to *Walter de Scoteney*, a Baron, who had also the maner of Hersham. But, for giving poison to Richard Earl of Clare, whose Steward he was, and to William, his brother, of which the latter died, he was hanged in 1259; and his estate, most probably, given to *John de Berewyk*, who died in 1312; holding the the maner of Toppesfield, of Gilbert de Clare, Earl of Gloucester, by the service of one Knight's-fee; and his heir was *Roger*, son of John *Huse*; more particularly mentioned under the maner of Husees. From him it came to *Tho. Rykedon*;

---

nence, 280 ft. above the sea. The *Church* (St. Margaret) is of brick, and has a nave, S. aisle of four bays, chancel, and embattled brick tower with 4 corner pinnacles and 5 bells; 3 dated 1675; one 1720; and one 1779. The body was built in 1519, the tower in 1699. In the chancel are mural tablets to Dorcas Smyth (1633); Robert Wildes (1690), rector; Thomas Willitt (1731), rector; the Rev. George Pawson (1797); and Elizabeth Erle (1655); also an uninscribed altar-tomb, on the S. side of the chancel, with floriated cross, probably to the founder of the church; and brasses to Wm. Cracherod, gent. (1585), and wife; and to John Cracherod (1534), and wife. There is also a fine incised stone, with an effigy of a cross-legged knight in armour, and a 14th century inscription to Thomas le Despenser. In the chancel is a piscina and another in the nave. The font is a rude, ancient one. The registers date from 1559. The women and children in this parish are partially engaged in straw-plaiting. *Essex (Co.) Handbook, by Miller Christy. London*, 1887.

TOPPESFIELD, ENGLAND.

The Parish Pump. An Old Resident.

The Winding Street.

St. Margaret's Tower. Berwick Hall.

and Robert Rykedon and others sold it, in 1420, to *John Doreward*, of Bocking, Esq., who, at the time of his decease, in the said year, held the maners and other lands, &c., called Berewyk, Scoteneys, and Cardeaux, in Toppesfield, the two Yeldhams, Mapiltrested, Haverill, Hengham Sible, and elsewhere. *John*, his son, succeeded him; and held this maner, with the lands, tenements, rents, and services, called Berwykes, Scoteneys, and Cardeaux, that composed the maner of Toppesfield, of Cecily, Duchess of York, as of her maner of Stamburne. He died in 1476. *John Doreward*, of Great Yeldham, Esq., held the same at the time of his death, the last day of February 1496; and *Christian*, his neice, brought it, in marriage, to her husband, *John de Vere*, the 14th Earl of Oxford on whom it was settled, in case of failure of issue, and on his heirs forever. In this noble family it continued, till Edward [the 17th] Earl of Oxford sold it [he having squandered away his various estates] 1st October 1584, to *William Bigge*, of Redgewell; who died possessed of it, 5th January 1585, and of Gounces, Brownes Farm, Broad-oake, with other estates adjoining. By his wife, Dorcas, daughter of John Mooteham, of this parish, Gent.,* he had William, Samuel, *Edward*, and Dorcas. *William*, the eldest son, who lived at Redfens in Shalford, held several parcels of land in this parish, belonging to the adjoining estate of Gunces; but *Edward*, the younger son had the maners of Berwick-hall and Scoteneys. *Edward*, his son, kept his first Court here on the 8th of October 1635.

In 1645, it came into the possession of *Robert Jacob*, Gent, and, in 1651, into that of *John Blackmore*, Esq. On the 23d of April 1658, *Robert Wankford*, Esq., kept his first Court here. He had two daughters by his first wife; and by his second; Robert, baptized 12th June 1631; and Samuel, 18th December 1632. *Robert*, his eldest son, seated at Berwick-hall, married Elizabeth, daughter and heir of Thomas Shelley, of Magdalen-Lavor in this county; and had by her, Berwick,

---

*She was remarried to William Smyth, of Cressing-Temple, Esq. and dying 18th December 1633, was buried at Cressing. But her grandson, Henry Bigge, Esq. erected a curious monument to her memory in the chancel of St. Margarets.

For illustration of this tablet, see, *The Ancient Sepulchral Monuments of Essex. By Frederic Chancellor*, p. 325, *London*, 1890.

who died young; *Robert*, Walter, Shelley; and seven daughters; of whom, Anne was married to John Elliston of Overhall in Gestingthorp, and afterwards to George Gent, Esq. Mary was wife of John Littel, of London, druggist; and the youngest, of Thomas Todd, of Sturmere. He died in 1688. *Robert*, his eldest surviving son, had no issue by his first wife, Dorothy, daughter of John Fotherby, of Rickmansworth in Hertfordshire, Esq.; but by his second wife, Mary, daughter of the Rev. John Oseley, Rector of Pantfeild, &c., he had several children. He was buried here on the 20th of June, 1708.

Some time after, the maners and demesnes of Berwicks, Scoteneys, and Gaynesfords, coming into the hands of Mr. John Poultnor, Attorney at Law, at Clare, he sold them to *Isaac Helbutt*, a rich merchant; from whom they passed to *Moses Hart*, and to *Wulph Ridolphus*, or, as some call him, *Michael Adolphus*, Esq.

THE MANER OF GAYNESFORDS,

Just now mentioned, took its name from an ancient family, who had also Gobions in this parish, Ashwell-hall in Finchingfield, Nicholls in Shaldford, &c. *Richard Gaynford*, who died 20th May 1484, held lands in this parish, which we suppose to be these. His brother John succeeded him. *William Butcher* held this capital messuage, and 24 acres of land, in Queen Elizabeth's reign. June 14, 1669, Thomas Guyver, with Samuel Edwards and Margaret his wife, daughter of Francis Guyver, sold this capital messuage to *Robert Wankford*; from whom they passed as above. Gaynesfords is near two miles south-west from the church.

THE MANER OF HUSEES.

Roger, son of *John Huse*, upon the death of John de Berewyk in 1312, inherited this estate, to which he gave name. This Roger sprung from the ancient family of Huse in Wiltshire and Dorsetshire; was a great soldier; became a knight; had summons to Parliament in 1348 and 1349, and died in 1361; being seated at Barton Stacy, in Hampshire. *John*, his son, succeeded him. In 1419, Alexander

Eustace and John Wood sold this estate to John Symonds. *Henry Parker*, of Gosfeild, Esq. who died 15th January 1541, held this messuage, called Hosees, and 80 acres of arable and meadow, of John de Vere, Earl of Oxford, in socage; besides other parcels here,* and great estates elsewhere. *Roger*, his son, succeeded him. *William Cratchrode*, junior, held this maner in 1585. About the latter end of Queen Elizabeth, it was holden by *John Alston*, of Belchamp Oton, who gave it to his third son, *Matthew*; and and he having no issue, bequeathed it to *Thomas Cracherode*; of whom it was purchased by Colonel *Stephen Piper*; and it is now in the possession of Dr. *Piper* [whose family sold it to Henry Sperling, Esq., of Dines Hall].

THE MANER OF CUST-HALL.

The mansion-house stands near a mile south-west form the church. It took its name from an ancient and considerable family† which were seated here in King Edward the Third's reign. Afterwards, it became the Cracherode family that had long been settled at a place called from them Cracherodes, in this parish. The first of the name that hath occurred to us, was *John Cracherode*, witness to a deed, 17th Richard II. 1393. His son *Robert*, was father of *John*, an Esquire under John de Vere, Earl of Oxford, at the battle of Azincourt. *John Cracherode*, Gent., son of the latter, married Agnes, daughter and heir of Sir John Gates, of Rivenhall; and had by her, *John*; William, Clerk of the Green Cloth to King Henry VIII, and Thomas, who had to wife Brigett, daughter of Aubrey de Vere, second son to John the 15th Earl of Oxford. *John*, the eldest son, paid ingress fine for Cust-hall in 1504. He married Agnes, daughter of Tho. Carter; and departing this life in 1534, was buried in the middle of this church, under a grave-stone,

*Namely, Shoremeadow, Foxholes; a messuage, called Dudmans, and 70 acres of arable and meadow; two tenements, called Griggs and Algers; St. John's Land, &c.

†The Cust family was originally of Yorkshire, but long seated in Lincolnshire; as may be seen in the Baronetage, vol. iv, p. 629, under the article of the Right Hon. Sir *John Cust*, present Speaker of the House of Commons.

with an inscription. They had four sons and four daughters; viz., Helen, wife of William Hunt, of Gosfeild, Gent.; Joan, of John Tendring, of Boreham, Gent.; Julian, of . . . Lee; and Jane, of Peter Fitch, of Writtle, Gent. *William*, the only son whose name is recorded, married Elizabeth, daughter of John Ray, of Denston in Suffolk. They lived 56 years together in wedlock. At the time of his decease, 10th January, 1585, he held this capital messuage, called Custs, and 20 acres of free land, belonging of old thereto; also a messuage, anciently called Cracherodes, and afterwards Colman's, in this parish and in Hedingham Sible; with several other parcels of land; particularly Albegeons, and Camois Parke, Pipers Pond, &c. He, and his wife, which died 17th February 1587, lie both buried in the chancel of this church, under a blue marble stone. They had issue five sons and one daughter; viz., Thomas; Matthew, of Cavendish; John, Charles, William. The daughter, named Anne, was wife of John Mootham.—*Thomas*, the eldest son, married Anne, daughter of Robert Mordaunt, of Hemstead in this county, Esq., a younger branch of the Lord Mordaunt, of Turvey in Bedfordshire; by whom he had William, who died without issue; Thomas; and four daughters: Frances, married to Robert Wilkins, of Bumsted; Anne, to John Alston, of Belchamp-Oton; Elizabeth, to John Fryer, of Paul's-Belchamp, and Barbara, to . . . . Harris. He died 14th June 1619.—*Thomas*, his son and heir, then aged 40 years, married Elizabeth, daughter of Richard Godbolt, of Finchamp in Norfolk; John, of Cranham-hall in Romford; Richard; and three daughters: Elizabeth, Brigett, and Susan.—*Mordaunt*, the eldest son, married Dorothy, daughter of Antony Sammes, of Hatfeild-Peverell. He died 2d of February 1666, and she 6th of March 1692. Both lie buried in this church.—They had issue, Thomas, baptized on the 17th of September 1646; Antony; Mordaunt [who was a linen-draper of London]; and Mary, wife of Christopher Layer, of Boughton-hall, Esq. Thomas, the eldest son, married Anne, daughter of Christopher Layer, of Belchamp St. Paul; by whom he had Thomas, baptized the 1st of June 1680. He was buried in this church the 8th of July 1706. *Thomas*, his son and heir, sold this maner, in 1708, to

Colonel *Stephen Piper*, mentioned a little before [whose family sold in to Henry Sperling, Esq., of Dines Hall].

THE MANER OF CAMOYS,

Is the largest in this parish; consisting, in time past, of two Knight's-fees, holden in the honor of Clare. The mansion-house stands near the church, and formerly had a park. In Edward the Confessor's reign, Got held this lordship, as lying in this parish and Stanburne, and then in two maners; which, at the time of the survey, belonged to Hamo Dapifer. How long it continued united with Stamborne, we cannot certainly discover.

Sir Ralph de Camoys,* from whom it borrowed its name, held it under Richard de Clare, Earl of Gloucester and Hertford, in 1262, as two Knight's-fees. He was a man of great note in his time; and after the taking of King Henry III, prisoner at the battle of Lewes, was chosen, by the discontented Barons, one of their Council of State, to govern the Realm.† He was also summoned to Parliament, 24th December 1264. He died in 1276.—*John*,‡ his son and successor, was father of *Ralph*, who gave this estate, in free-marriage with his daughter *Ela*, to *Peter Gonsell*, or Gonshill. This family was originally of Yorkshire, *Giles Gonsell*, by Eminentia, daughter of Fulk de Oyry, of Gedney in Lincolnshire, had *Peter*; who, by the said Ela his wife, had *Ralph* and Margaret. Ralph dying in 1295, was succeeded by his sister, *Margaret*, who had two husbands, first, *Philip le Despenser*, 4th son of Hugh le Despenser, Earl of Gloucester. He

*The name of Cammois is in the list of those that came in with William the Conqueror.—*Chronic. J. Bromton, col.* 963.

†See *Dugdale's Baron. vol. i, p.* 767.

‡This John married Margaret, daughter and heir of Sir John de Gatesden; and she forsaking him, and living in adultery with Sir William Paynel, John de Cameys, as he calls himself, quitted all his right and title to her, as also to all her goods and chattels, spontaneously delivering and demising her unto the said Sir William, and releasing all title and claim to her and her appertenances; as appears by the deed, printed at length in Sir William Dugdale's Baron. vol. i, p. 767.—After her lawful husband's decease, she was married to the said Sir William, and claimed thirds of Camoys estate; which the Parliament, out of due regard to morality and law, refused her.

departing this life in 1313, she took to her second husband, Sir John Roos, and lived till 1349. By her first husband, she had *Philip le Despenser;* who, at the time of his decease, in 1349, jointly with Joane his wife, held, of the Lady of Clare, a tenement here called Camoy's-hall, by the service aforesaid. *Philip*, his son, by . . . daughter of . . . Strange, had Philip, who died in 1400; leaving, by his wife, Margaret Cobham, Sir *Philip*, his son and heir, that departed this life in 1423, and held this maner of Edward, Earl of March; as also those of Lyndsells, Little Stambridge, and a fourth part of the maner of Thaxted. He married Elizabeth, one of the daughters and coheirs of Sir Robert Tiptoft; and by her he had his only daughter and heir, *Margery*. She was married, first, to Sir *Roger Wentworth*, third son of John Wentworth, of Elmes-hall in Yorkshire, Esq. a younger branch of the Wentworths, of Wentworth Woodhouse; from whence are descended the Earls of Stratford. Her second husband was John Lord Rosse; by whom she had no issue. But by her first husband, she had two sons; Philip; and Henry, the first of this family seated at Codham-hall; from whom sprung the Wentworths, of Gosfeild and Bocking; and several daughters. She died the 20th of April 1475. Sir *Philip Wentworth*, her eldest son, and heir to this estate, married Mary, daughter of John Lord Clifford; and had by her, Sir *Henry*, father of Sir *Richard*, a Knight-Banneret; who, by Anne, daughter of Sir James Tyrell, of Gipping in Suffolk, had Sir *Thomas* Wentworth, of Nettlested, created Baron Wentworth the 2d of December 1529. He married Margaret, daughter of Sir Adrian Fortescue; and had by her, *Thomas*, Lord *Wentworth*, who held his first Court here the 16th of June 1551.—He had also the maners Hackney and Stepney; and was the last Governor of Calais under Queen Mary I. The 4th of 13th of May 1557, he sold Camoys-hall to *William Fitch*, Esquire, of Little Canfield It continued little more than twenty years in his name, for he dying the 20th of December 1578, it came to his son Thomas; who surviving him but a little while, it then fell to his only daughter and heir, *Mary*, that had been married, about the year 1556, to *Francis Mannock*, Esq. . . . . who died 3d of November 1590 and was succeeded by his son *William*; whose son and heir, *Francis*, was created a Baronet the 1st of June 1627; and had for successors, Sir *Francis* and Sir *William*. The latter sold this estate, the 25th of March 1713, to *Matthias Unwin*, of Castle Hedingham, Gent, who died the 18th of September 1715; and, by will, bequeathed Camoys-hall to his brother's son, *Joseph*. This latter dying in September 1747, was succeeded by his eldest son, *Joseph Unwin*, Gent. [of Castle Hedingham.]

FLOWERS-HALL,

Is about two miles south south-west from the church. From a family that existed here from 1369 to 1572, it took the name of Flowers. *Thomas Glascock*, who died 29th October 1631, held the maner and capital messuage called Flowers-hall, Giddings, and Brownes, with appertenances, of Edward Benlowes, Esq, of his maner of Justices, in Finchingfield, by the annual rent of 8 s. one cock, one hen, and an egg and a half. It was afterwards *Henry Glascocks*.* This estate paid quit-rent to Nortofts in Finchingfeild.

GOBIONS,

Is denominated from an ancient knightly family, surnamed Gobyon, that had considerable estates at Finchingfeild, Bardfeild, Great Lees, Laindow, East Tilbury, &c. . . . . Sir Thomas Gobion was High Sheriff of Essex and Hertfordshire in 1323. . . . John Gobyon is in the list of the gentry of this county in 1433. *Richard Gainford*, mentioned above, under Gaynesfords, held this maner of Gobyns in 1483, of John Doreward, as of his maner of Great Yeldham. *John*, his brother, was his heir. It was afterwards in the Wentworth family.

HAWKES-HALL,

Formerly belonged to a family surnamed De Hausted; from whom it passed to the St. Martins, and the noble family of Bourchier; in which last it continued long. Some of their

*This estate afterwards became the property of Mr. Ralph Jephson, by marriage with the daughter of William Raymond, of Notley.

mesne or under-tenants were, Joane, daughter of John Gilderich, of Peches in Finchingfeild, about 1422; and John Helyoun, Lord of the maner of Bumstead-Helion, in 1450. It is described as comprehending 100 acres of arable, 8 acres of meadow, 8 acres of pasture, and 10 acres of wood. It passed since to Bendlowes, &c., as Justices in Finchingfeild.

THE MANER OF BRADFIELD,

Near a mile sout-west from the church, was holden, about the year 1393, by John Bradfend or Bradfeild, from whom it received its name. He had also the maner of Nicholls in Shalford. William Toppesfeild held it of John Durward, at the time of his decease, in 1480; and his two daughters, Elizabeth and *Joane* Toppesfeild, were his heirs. The latter brought it in marriage to . . . *Paynell*, and was his widow in 1498. The Paynell, or Pannell family, was in these parts as early as the reign of King Edward I, and had an estate at Redgewell, where John Pannell lived in 1385, and his posterity continued till the reign of King James I. *Henry Pannell*, Esq., who died the 18th of July 1573, held this maner of Bradfield of the Earl of Oxford, as of his maner of Berewikes, and other lands here. His son and heir, *Henry*, was then 12 years old. [This estate afterwards passed into the hands of Mr. John Darby, of Little Waltham, Essex co., and at his death devolved to Mr. Solomon Edwards of Thackstead.]*

*Some curious Roman remains were found on June 28, 1800, by a labourer making a ditch at the bottom of Red Bamfield, belonging to Bradfield Farm, situate about two miles west by south of the ancient Roman road from Camulodunum to Camboritum, (Colchester to Cambridge).

"The sword blade, which was very much corroded and broken in two or three places, lay across the breast of the skeleton found therewith; it was rather a singular situation, for in general they are found by the side of the person interred.

The metal vase and *patera* merit attention. The vase was of that form which Montfaucon calls a *preæfericulum* used by the Romans at their sacrifices for pouring wine into the *patera*.

The uses of the elegant little cups of Samian ware, one of which has an ornamented border, have not, that I can find, been ascertained. As they were interred with the corpse we may suppose them to have contained holy oil, gums, balsams, unguents, &c., but this is conjecture only. The real purposes to which they were applied must remain at present in obscurity; we only know that such things were

AN OLD COTTAGE, TOPPESFIELD, ENGLAND.

THE RECTORY, TOPPESFIELD, ENGLAND.

OLIVERS is an ancient capital messuage in this parish, about three quarters of a mile south-east from the church. John Oliver purchased an estate of John de Raclesden, about 1360, which is supposed to have been this. He was one of Sir. John Hawkwood's Esquires, companions, and fellow-warriors; and concerned in founding his Chantry.*

Richard Simon was possessed, in 1627, of this tenement, called Olivers and Dudmans, and, in 1631, Thomas Glascock, above mentioned, had a messuage, and 12 acres of land thereto belonging, called Olivers; † with Ashleies and Gadleies, two other parcels. Here were in this parish two acres and a half of land, called Molle, given for one obit and a lamp;

---

used at their funeral obsequies, particularly unguents and perfumes of several kinds for anointing the body before interment; therefore we may conclude that they were used at the funeral, and were afterwards deposited with the body, according to the custom of the ancients.

Only one Roman coin was found, and that very imperfect. Whether it was the *obolus*, the *naulum Charontis*, is left for others to determine. A nail and a handle of a bronze *patera* were found at the same time."—*Archæologia, vol. xiv, pp.* 24–26, 2 *plates*, *London*, 1803.

*The friends and executors of Sir. John Hawkwood founded a famous *chantry*, for one Chaplain in the church of Hedingham, to pray for the souls of Sir John Hawkwood, Thomas Oliver, and John Newenton, Esquires, his military companions, supposed to be born in this county. The license for this foundation, was in 1412; and the endowment consisted of 4 messuages, 4 tofts, 420 acres of arable, 13 acres of meadow, 20 of pasture, 4 of wood, 22 of alder, and 12 s. rent, in Sible and Castle Hengham, Gosfeild, Mapiltrested, Great and Little Gelham, and Toppesfeild. The house where the Chantry Priest lived stands at some distance from the church, and bore then, and still bears, the name of Hostage; having originally been a charitable foundation for the entertainment of devout Pilgrims. The patronage of this chantry belonged to the Lord of the maner of Hawkwoods.

†This estate was occupied at one time, by Samuel Symonds, gent., who came to New England, in 1637, and settled at Ipswich, where the town granted him a farm of five hundred acres, lying partly within the present bounds of Topsfield. This farm was known on the records as "Olivers." *See ante, pp.* 40, 41.

The family of Symonds was originally of Croft in Lancashire, where they continued in a direct line for about twenty generations. Richard Symonds of the third generation was seated in Great Yeldham, at "The Pool," on the eastern bank of the river Colne. He married, Jan. 9, 1580, Elizabeth, daughter of Robert Plumb, of Yeldham Hall. Samuel, the third son, married Dorothy, daughter of Thomas Harlakenden, of Earl's-Colne; was a Cursitor in Chancery; and had Oliver's in Toppesfield; but retired to New England with his family. *Morant.*

with about three acres more; which, at the suppression of Chantries, were granted to Thomas Golding, Esq. Samuel Hurrell, John Piper, Geffrey Cook, Matthias and Edmund Davey, Tho. Orford, and Tho. Teader, have also estates here. This parish is rated to the land-tax at 1692 £. 1 s. 4 d.

The CHURCH, dedicated to St. Margaret, is tolerably handsome and spacious. It was formerly, all leaded; but is now only so in part. The chancel is tiled. About 70 years ago, the tower, which was built of flint and stone fell down; but hath since been rebuilt, of brick, in a firm and substantial manner; towards which, Mr. Wilde, Rector at the time it fell, bequeathed 100 £. To it belong five bells. Here was, anciently, a rectory and vicarage; of both which, the Prior and Convent of Stoke near Clare, whilst a priory, and when a college, the Dean and Chapter, were patrons. In what year, and by whom given to them, is unknown. The rectory was a sinecure; and so continued, till Thomas Kemp, Bishop of London, finding the vicarage was grown so poor* that it could not maintain a Vicar, or discharge the burdens incumbent thereon, so that it had been vacant and neglected several years, he reunited and incorporated again the rectory and vicarage. At the dissolution of religious houses, the patronage of this rectory coming to the Crown, King Edward VI. gave it to his proeceptor, Sir John Cheke; upon whose unhappy fall, it reverted to the Crown, and hath remained in it ever since; it being a considerable living. There are lands of about six pounds a year, belonging to the church.

TOPPESFIELD, ENG. * * * "I found the ride exceedingly pleasant, along the narrow but excellent road, which winds its way through an unbroken succession of luxuriant cornfields and meadows. * * * It was evening when I arrived, and the 'Green Man Inn' received me. This is a small, but neat and comfortable tavern, and bears the marks

*At the petition of William Parker then rector, with the consent of the dean and chapter of St. Paul, and the arch-deacon of Middlesex.

of a respectable antiquity. It is, in fact, just such a place as the ale-house of Goldsmith's poem, and has been, I presume, the nightly resort of the Toppesfield politicians, for at least two hundred years.

When I went out the next morning, I found myself in a small village, composed of stone cottages, mostly plastered, white-washed and thatched. I saw nothing in them particularly pleasing, beyond that aspect of neatness, and those floral adornings, which rarely desert even the meanest rural home in that beautiful country. My first visit was to the church of St. Margaret. * * * The interior interested me much. A place of worship more rude in aspect, or less adapted to comfort, it would, I am sure, be difficult to find in all New England. * * * The pews are narrow, upright boxes, with high sides, and, with the exception of the Rector's, are uncushioned and uncarpeted, a few of them, however, were supplied with straw covered hassocks. Upon the southern side there are four Gothic arches, which rest upon short thick columns. On this side there is a low gallery, erected, as an inscription shows, in 1833. The pulpit and reading desk are on the opposite side. These are of oak, and the former resembles, in shape and appearance, that interesting relic, the old Capen pulpit. * * * [In the church registers I found] the name of Samuel Symonds, gent., and that of Dorothy his wife. Between 1621 and 1633, I found and copied the baptisms of ten of their children. * * * The Parsonage is a charming residence, surrounded by flowers and shrubbery, and smooth-shaven lawns. The present incumbent lives among his people and seems to be regarded with respect and affection. * * * Here I was in a community of several hundred people, not a man of whom owns one rood of the land which he cultivates—not an individual of whom possesses the house that shelters him. These skillful farmers are tenants at will—and are perpetually struggling under an oppressive burden of rents, and tythes, and taxes, and rates. These hardy laborers think they do well, if their toil yields them the average remuneration of a shilling a day. As to religious privileges they have indeed a sitting, hired or free, in yonder rude church. Their Rector, sent them by the Queen, may be a good man, or he may not. With the question of his appointment or dismission, they have just as much concern as you have. They are, however, permitted to pay him. From that glebe, which is made so rich by their sweat, he draws an annual stipend, three times as large as that which you raise for your two clergymen. And here, in a parish which pays its Rector more than thirty-five hundred dollars a year,—here within four hours ride of the grand metropolis of the world, here, in the middle of the nineteenth century, a free school is a thing which yet remains to be invented."—*Nehemiah Cleaveland, in Salem Register, Nov. 1851.*

TOPPESFIELD, ENG. * * * "At Yeldham the only cab we could find was a little dog-cart with a Welch pony that hardly came up to the shafts. However, this was all that was necessary and the owner told us he would take us for two shillings if we 'didn't think that much would harm us.' He proved himself capable of giving considerable information about the church and the chapels (as Congregational and Methodist churches are called in England) as his father had been Parish Clerk at Yeldham for a good many years, but when I asked him the origin of the name Toppesfield his answer was: 'Well, that's a question I could hardly answer, Sir. They must-a-caught it as it come along. Come by a whirlwind perhaps.' Mr. Lane, the genial teacher of the parish, told us that the only reason he could find was from the fact of its being the topmost village in the shire. * * * We had been informed that some years before, a gentleman from Topsfield, America, had come to see the graves of his ancestors; the woman who told us could not remember the name, and so we mentioned over the names of Cleveland, Peabody, Bradstreet, thinking it might be some of these, but none of them seemed familiar. Finally the mother came in and said: 'Why, it was the one who had six wives, Joseph Smith* was the name.'

*Joseph Smith, the Mormon prophet, was of Topsfield ancestry. The Smith referred to may have been a descendant.

The present school was built in 1856 by the then Rector, Rev. Mr. Gooch. It has accommodations for two hundred children and has one hundred and forty names on the register."—*Rev. Lyndon S. Crawford, in Salem Gazette, Nov. 25, 1886.*

TOPPESFIELD, ENG. * * * "All the fields are separated by hedges and these are generally well kept. The whole country looks neat and tidy. * * * The village was but a handful of houses along a narrow road or two, without any sidewalks to speak of. We left our traps at the 'Green Man' inn and got a glass of home brew, rather sour and not very good. * * * The Church itself is not at all large, and would hardly seat two hundred persons. It was built early in the 16th century, and has been very well preserved, Even during the Commonwealth, it was not much disturbed. It is one of the very few parishes whose records are kept throughout that period without a break. We were assured that that was a very unusual circumstance."—*Brandreth Symonds, in Essex County Mercury, Oct. 24, 1894.*

TOPPESFIELD, ENG. * * * In approaching Toppesfield, the high hills of the town come into view before the train leaves you in the valley. The station building might be called a primitive one:—old, dilapidated, and inconvenient. Nevertheless it serves for the transaction of the limited business of a small country station. The village is about one and a half miles from the station, if one takes the short cut across the fields on foot in a direct line. The road makes a detour in a southerly and southwesterly and then in a northwesterly course to avoid the steep acclivity, and covers about two miles before reaching the village. The way for the most part is a gentle ascent,—one rise of many rods being steeper than the rest.

We first reach that part of the village where the rectory is located. It is large and commodious for a place of the size of that in which it is situated. The building is almost entirely obscured by shade trees, shrubbery and evergreen. Passing on some twenty or thirty rods, in a northerly direction, going by several dwellings we come to the end of the street that we have traversed. Here we meet another street lying east and west,—the principal street of the village. Near the right hand corner is St. Margaret's—the parish church. Farther on to the right is the school house. Near the left hand corner is a chapel where the Nonconformists worship. To the westward some rods, is the post-office.

I did not explore the whole village, but it will be seen by the location of the public buildings that I was in the central and most important part of it. St. Margaret's Church has been an active force in the village for eight hundred years. . . . The interior as well as the exterior has all the marks of an old structure. Few changes have been made in modern times that conceal its ancient appearance. * * * A tablet on the wall of the interior has a list of rectors extending back three hundred years and more, I transcribed some of the names that may be interesting to Topsfield people. 1559, Thomas Donnell, B. D.; 1601, William Smith; 1604, Randolph Davenport; 1662, Nathaniel Ward; 1691, Thomas Willett; 1694, Robert Wilde.

A curious fact to be noticed in the list of rectors is that in the days of the Commonwealth there is a break in the list with a statement that there was a vacancy in those years. Although there was no "rector," doubtless there was preaching in the church by Dissenters in that interval. The church stands in the midst of, and is entirely surrounded by the churchyard. The small cemetery is still in use for burials. I noticed that they were opening graves in what appeared to be the oldest part of the yard. The inscriptions on the oldest monuments are illegible as well they might be in a cemetery eight hundred years old. I noticed the monument of Henry Howlett, who died in 1773, aged 72.

The chapel of the Nonconformists I did not enter. It is a very plain and unpretending building.

The post office is in the house of the post master. Apartments of modest proportions are set apart for the government office. There is no room for the floating population of the town to assemble in for social intercourse, to talk over the news of the day, and enjoy the village gossip. In fact if there was such a place in the village I failed to discover it.

The houses, barns, and out-buildings are generally built of brick. The style of architecture is not pretentious. There is not the facility for architectural display in small brick buildings, that there is by working in wood. I noticed here as well as through England, as far as I travelled, the projecting second story of old houses, like that of our own Capen house. One house in particular, better than the average, in the old style, I was informed was a modern built house. They have a way in England, and I think to a great extent, of building after the style of several hundred years ago, to have the buildings conform those in the neighborhood.

The most of the people, I suppose, would be reckoned in the middle class. Some as indigent or poor. The better classes have comfortable homes, and show intelligence and cultivation.

Toppesfield is especially an agricultural town. It has a good soil. The soil of Essex is not as fertile as that of some other parts of the kingdom. I heard Englishmen in speaking of the county, say that the land in Essex is poor. Such may be the case as far as the county in general is considered, but I think an exception must be made in the case of the plateau upon which Toppesfield is situated; for there the farmers were harvesting good crops and the land was making abundant returns for the labor and skill of the husbandmen; much better probably than the average of the county. The principal crops are wheat, barley, vegetables and hay. Being remote from any large town, market gardening is not carried on. Much of the hay crop is stacked in the fields where it is gathered, as it is in other parts of England. I noticed stacks that had breasted the storms of one or more winters, notwithstanding the great demand for forage on account of the wars in which the nation was engaged. The barley product is largely used for malt to brew the universal English beer. It was wheat harvest when I was there. I saw an abundant yield of wheat on the highest land in the village, as large, I should judge, as that of the most fertile parts of the island. The parish of St. Margeret's has some of the best land in the place, I do not know how many acres, some of which is divided into small "allotments," each of an acre or less, one half, one quarter, or one eighth of an acre. These are let, at a low rental, to indigent people of the parish who have no land, the proceeds of which go to help other poor people.

The following Toppesfield names taken from the voting list are of interest as being common to our own Topsfield and vicinity:—Allen, Barker, Barnes, Clarke, Davison, Hale, Hardy, Palmer, Reed, Rice, Smith, Wilson.

Justin Allen, M. D., March 15, 1901.

www.ingramcontent.com/pod-product-compliance
Ingram Content Group UK Ltd.
Pitfield, Milton Keynes, MK11 3LW, UK
UKHW042014190726
13854UKWH00005B/2283

9 789351 287766